Cracking the Data Engineering Interview

Land your dream job with the help of resume-building tips, over 100 mock questions, and a unique portfolio

Kedeisha Bryan

Taamir Ransome

BIRMINGHAM—MUMBAI

Cracking the Data Engineering Interview

Group Product Manager: Kaustubh Manglurkar

Publishing Product Manager: Arindam Majumder

Book Project Manager: Farheen Fatima

Senior Editor: Nathanya Dias

Technical Editor: Sweety Pagaria

Copy Editor: Safis Editing

Proofreader: Safis Editing

Indexer: Hemangini Bari

Production Designer: Vijay Kamble

DevRel Marketing Coordinator: Nivedita Singh

First published: November 2023

Production reference: 1261023

Published by Packt Publishing Ltd.

Grosvenor House

11 St Paul's Square

Birmingham

B3 1RB, UK.

ISBN 978-1-83763-077-6

www.packtpub.com

To my father, Owen Bryan Sr, who has been a rock in my corner in all my endeavors. And always reminding me of my talents when I can't see them myself.

– Kedeisha Bryan

Thanks

-Taamir Ransome

Contributors

About the authors

Kedeisha Bryan is a data professional with experience in data analytics, science, and engineering. She has prior experience combining both Six Sigma and analytics to provide data solutions that have impacted policy changes and leadership decisions. She is fluent in tools such as SQL, Python, and Tableau.

She is the founder and leader at the Data in Motion Academy, providing personalized skill development, resources, and training at scale to aspiring data professionals across the globe. Her other works include another Packt book in the works and an SQL course for LinkedIn Learning.

Taamir Ransome is a Data Scientist and Software Engineer. He has experience in building machine learning and artificial intelligence solutions for the US Army. He is also the founder of the Vet Dev Institute, where he currently provides cloud-based data solutions for clients. He holds a master's degree in Analytics from Western Governors University.

About the reviewers

Hakeem Lawrence is a highly skilled Power BI analyst with a deep passion for data-driven insights. He has mastered the art of transforming complex datasets into compelling visual narratives. However, his expertise extends beyond Power BI; he is also a proficient Python developer, adept at leveraging its data manipulation and analysis libraries. His analytical prowess and coding finesse have enabled him to create end-to-end data solutions that empower organizations to make informed decisions. He is also a technical reviewer for Kedeisha Bryan's second book, *Becoming a Data Analyst*.

Sanghamitra Bhattacharjee is a Data Engineering Leader at Meta and was previously Director of Machine Learning Platforms at NatWest. She has led global transformation initiatives in the Data and Analytics domain over the last 20 years. Her notable work includes contributions to mobile analytics, personalization, real-time user reach, and NLP products.

She is extremely passionate about diversity and inclusion at work and is a core member of Grace Hopper Celebrations, India. She has organized conferences and meet-ups and she has been a speaker at several international and national conferences, including NASSCOM GCC Conclave, Microstrategy World, and Agile India. She was also awarded a patent for her work on delivering contextual ads for search engines.

Abhishek Mittal is a Data Engineering & Analytics professional with over 10 years of experience in business intelligence and data warehousing space. He delivers exceptional value to his customers by designing high-quality solutions and leading their successful implementations. His work entails architecting solutions for complex data problems for various clients across various business domains, managing technical scope and client expectations, and managing implementations of the solution. He is a Microsoft Azure, Power BI, Power Platform, and Snowflake-certified professional and works as a Principal Architect with Nagarro. He is also a Microsoft Certified Trainer and is deeply passionate about continuous learning and exploring new skills.

Table of Contents

6

Unit Testing   63

7

Database Fundamentals   71

8

Essential SQL for Data Engineers   81

Part 3: Essentials for Data Engineers Part II

9

10

11

12

Part 4: Essentials for Data Engineers Part III

13

14

15

Data Security and Privacy   151

16

Additional Interview Questions   159

Index   165

Other Books You May Enjoy   178

Preface

Within the domain of data, a distinct group of experts known as data engineers are devoted to ensuring that data is not merely accumulated, but rather refined, dependable, and prepared for analysis. Due to the emergence of big data technologies and the development of data-driven decision-making, the significance of this position has increased substantially, rendering data engineering one of the most desirable careers in the technology sector. However, the trajectory toward becoming a prosperous data engineer remains obscure for many.

Cracking the Data Engineering Interview serves as a printed mentor. Providing ambitious data engineers with the necessary information, tactics, and self-assurance to enter this ever-changing industry. The organization of this book facilitates your progression in comprehending the domain of data engineering, attaining proficiency in its fundamental principles, and equipping yourself to confront the intricacies of its interviews.

Part 1 of this book delves into the functions and obligations of a data engineer and offers advice on establishing a favorable impression before the interview. This includes strategies, such as presenting portfolio projects and enhancing one's LinkedIn profile. *Parts 2 and 3* are devoted to the technical fundamentals, guaranteeing that you will possess a comprehensive understanding of the essential competencies and domains of knowledge, ranging from the intricacies of data warehouses and data lakes to Python programming. In *Part 4*, an examination is conducted of the essential tools and methodologies that are critical in the contemporary data engineering domain. Additionally, a curated compilation of interview inquiries is provided for review.

Who this book is for

If you are an aspiring Data Engineer looking for a guide on how to land, prepare, and excel in data engineering interviews, then this book is for you.

You should already understand and should have been exposed to fundamentals of Data Engineering such as data modeling, cloud warehouses, programming (python & SQL), building data pipelines, scheduling your workflows (Airflow), and APIs.

What this book covers

Chapter 1, The Roles and Responsibilities of a Data Engineer, explores the complex array of responsibilities that comprise the core of a data engineer's role. This chapter unifies the daily responsibilities, long-term projects, and collaborative obligations associated with the title, thereby offering a comprehensive perspective of the profession.

Chapter 2, Must-Have Data Engineering Portfolio Projects, this chapter helps you dive deep into a selection of key projects that can showcase your prowess in data engineering, offering potential employers tangible proof of your capabilities.

Chapter 3, Building Your Data Engineering Brand on LinkedIn, this chapter shows you how to make the most of LinkedIn to show off your accomplishments, skills, and goals in the field of data engineering.

Chapter 4, Preparing for Behavioral Interviews, Along with technical skills, the most important thing is that you can fit in with your team and the company's culture. There are tips in this chapter on how to do well in behavioral interviews so that you can talk about your strengths and values clearly.

Chapter 5, Essential Python for Data Engineers, Python is still an important tool for data engineers. This chapter will help you learn about the Python ideas, libraries, and patterns that every data engineer needs to know.

Chapter 6, Unit Testing, In data engineering, quality assurance is a must. This chapter will teach you the basics of unit testing to make sure that your data processing scripts and pipelines are reliable and strong.

Chapter 7, Database Fundamentals, At the heart of data engineering lies the database. In this chapter you will acquaint yourself with the foundational concepts, types, and operations of databases, establishing a solid base for advanced topics.

Chapter 8, Essential SQL for Data Engineers, SQL is the standard language for working with data. This chapter will help you learn the ins and outs of SQL queries, optimizations, and best practices so that getting and changing data is easy.

Chapter 9, Database Design and Optimization, It's both an art and a science to make databases work well. This chapter will teach you about advanced design principles and optimization methods to make sure your databases are quick, scalable, and reliable.

Chapter 10, Data Processing and ETL, Turn raw data into insights that can be used. In this chapter we will learn about the tools, techniques, and best practices of data processing in this chapter, which is about the **Extract, Transform, Load** (ETL) process.

Chapter 11, Data Pipeline Design for Data Engineers, A data-driven organization needs to be able to easily move data from one place to another. In this chapter you will learn about the architecture, design, and upkeep of data pipelines to make sure that data moves quickly and reliably.

Chapter 12, Data Warehouses and Data Lakes, Explore the huge world of ways to store data. This chapter teaches you the differences between data warehouses and data lakes, as well as their uses and architectures, to be ready for the challenges of modern data.

Chapter 13, Essential Tools You Should Know About, It's important to have the right tool. In this chapter you will learn how to use the most important tools in the data engineering ecosystem, from importing data to managing it and keeping an eye on it.

Chapter 14, Continuous Integration/Continuous Development for Data Engineers, Being flexible is important in a world where data is always changing. In data engineering and in this chapter, you will learn how to use CI/CD to make sure that data pipelines and processes are always up-to-date and running at their best.

Chapter 15, Data Security and Privacy, It's important to be responsible when you have a lot of data. This chapter will teach you about the important issues of data security and privacy, and get to know the best ways to protect your data assets and the tools you can use to do so.

Chapter 16, Additional Interview Questions, Getting ready is half the battle won. This chapter comprises of carefully chosen set of interview questions that cover a wide range of topics, from technical to situational. This way, you'll be ready for any surprise that comes your way.

To get the most out of this book

You will need to have a basic understanding of Microsoft Azure.

Software/hardware covered in the book	Operating system requirements
Microsoft Azure	Windows, macOS, or Linux
Amazon Web Services	Windows, macOS, or Linux
Python	Windows, macOS, or Linux

Download the example code files

You can download the example code files for this book from GitHub at `https://github.com/PacktPublishing/Cracking-Data-Engineering-Interview-Guide`. If there's an update to the code, it will be updated in the GitHub repository.

We also have other code bundles from our rich catalog of books and videos available at `https://github.com/PacktPublishing/`. Check them out!

Conventions used

There are a number of text conventions used throughout this book.

`Code in text`: Indicates code words in text, database table names, folder names, filenames, file extensions, pathnames, dummy URLs, user input, and Twitter handles. Here is an example: "Mount the downloaded `WebStorm-10*.dmg` disk image file as another disk in your system."

A block of code is set as follows:

```
from scrape import *
import pandas as pd
from sqlalchemy import create_engine
import psycopg2
```

Bold: Indicates a new term, an important word, or words that you see onscreen. For instance, words in menus or dialog boxes appear in **bold**. Here is an example: "You can get your connection string from your **Connect** tab and fix it into the format shown previously."

> **Tips or important notes**
> Appear like this.

Get in touch

Feedback from our readers is always welcome.

General feedback: If you have questions about any aspect of this book, email us at `customercare@packtpub.com` and mention the book title in the subject of your message.

Errata: Although we have taken every care to ensure the accuracy of our content, mistakes do happen. If you have found a mistake in this book, we would be grateful if you would report this to us. Please visit `www.packtpub.com/support/errata` and fill in the form.

Piracy: If you come across any illegal copies of our works in any form on the internet, we would be grateful if you would provide us with the location address or website name. Please contact us at `copyright@packt.com` with a link to the material.

If you are interested in becoming an author: If there is a topic that you have expertise in and you are interested in either writing or contributing to a book, please visit `authors.packtpub.com`.

Share Your Thoughts

Once you've read *Cracking the Data Engineering Interview*, we'd love to hear your thoughts! Scan the QR code below to go straight to the Amazon review page for this book and share your feedback.

https://packt.link/r/1-837-63077-1

Your review is important to us and the tech community and will help us make sure we're delivering excellent quality content.

Download a free PDF copy of this book

Thanks for purchasing this book!

Do you like to read on the go but are unable to carry your print books everywhere?

Is your eBook purchase not compatible with the device of your choice?

Don't worry, now with every Packt book you get a DRM-free PDF version of that book at no cost.

Read anywhere, any place, on any device. Search, copy, and paste code from your favorite technical books directly into your application.

The perks don't stop there, you can get exclusive access to discounts, newsletters, and great free content in your inbox daily

Follow these simple steps to get the benefits:

1. Scan the QR code or visit the link below

https://packt.link/free-ebook/9781837630776

2. Submit your proof of purchase
3. That's it! We'll send your free PDF and other benefits to your email directly

Part 1: Landing Your First Data Engineering Job

In this part, we will focus on the different types of data engineers and how to best present yourself in your job hunt.

This part has the following chapters:

- *Chapter 1, The Roles and Responsibilities of a Data Engineer*
- *Chapter 2, Must-Have Data Engineering Portfolio Projects*
- *Chapter 3, Building Your Data Engineering Brand on LinkedIn*
- *Chapter 4, Preparing for Behavioral Interviews*

1

The Roles and Responsibilities of a Data Engineer

Gaining proficiency in data engineering requires you to grasp the subtleties of the field and become proficient in key technologies. The duties and responsibilities of a data engineer and the technology stack you should be familiar with are all explained in this chapter, which acts as your guide.

Data engineers are tasked with a broad range of duties because their work forms the foundation of an organization's data ecosystem. These duties include ensuring data security and quality as well as designing scalable data pipelines. The first step to succeeding in your interviews and landing a job involves being aware of what is expected of you in this role.

In this chapter, we will cover the following topics:

- Roles and responsibilities of a data engineer
- An overview of the data engineering tech stack

Roles and responsibilities of a data engineer

Data engineers are responsible for the design and maintenance of an organization's data infrastructure. In contrast to data scientists and data analysts, who focus on deriving insights from data and translating them into actionable business strategies, data engineers ensure that data is clean, reliable, and easily accessible.

Responsibilities

You will wear multiple hats as a data engineer, juggling various tasks crucial to the success of data-driven initiatives within an organization. Your responsibilities range from the technical complexities of data architecture to the interpersonal skills necessary for effective collaboration. Next, we explore the key responsibilities that define the role of a data engineer, giving you an understanding of what will be expected of you as a data engineer:

- **Data modeling and architecture**: The responsibility of a data engineer is to design data management systems. This entails designing the structure of databases, determining how data will be stored, accessed, and integrated across multiple sources, and implementing the design. Data engineers account for both the current and potential future data needs of an organization, ensuring scalability and efficiency.

- **Extract, Transform, Load (ETL)**: Data extraction from various sources, including structured databases and unstructured sources such as weblogs. Transforming this data into a usable form that may include enrichment, cleaning, and aggregations. Loading the transformed data into a data store.

- **Data quality and governance**: It is essential to ensure the accuracy, consistency, and security of data. Data engineers conduct quality checks to identify and rectify any data inconsistencies or errors. In addition, they play a crucial role in maintaining data privacy and compliance with applicable regulations, ensuring that data is reliable and legally sound.

- **Collaboration with data scientists, analysts, and other stakeholders**: Data engineers collaborate with data scientists to ensure they have the appropriate datasets and tools to conduct their analyses. In addition, they work with business analysts, product managers, and other stakeholders to comprehend their data requirements and deliver accordingly. Understanding the requirements of these stakeholders is essential to ensuring that the data infrastructure is both relevant and valuable.

In conclusion, the data engineer's role is multifaceted and bridges the gap between raw data sources and actionable business insights. Their work serves as the basis for data-driven decisions, playing a crucial role in the modern data ecosystem.

An overview of the data engineering tech stack

Mastering the appropriate set of tools and technologies is crucial for career success in the constantly evolving field of data engineering. At the core are programming languages such as Python, which is prized for its readability and rich ecosystem of data-centric libraries. Java is widely recognized for its robustness and scalability, particularly in enterprise environments. Scala, which is frequently employed alongside Apache Spark, offers functional programming capabilities and excels at real-time data processing tasks.

SQL databases such as Oracle, MySQL, and Microsoft SQL Server are examples of on-premise storage solutions for structured data. They provide querying capabilities and are a standard component of transactional applications. NoSQL databases, such as MongoDB, Cassandra, and Redis, offer the required scalability and flexibility for unstructured or semi-structured data. In addition, data lakes such as **Amazon Simple Storage Service** (**Amazon S3**) and **Azure Data Lake Storage** (**ADLS**) are popular cloud storage solutions.

Data processing frameworks are also an essential component of the technology stack. Apache Spark distinguishes itself as a fast, in-memory data processing engine with development APIs, which makes it ideal for big data tasks. Hadoop is a dependable option for batch processing large datasets and is frequently combined with other tools such as Hive and Pig. Apache Airflow satisfies this need with its programmatic scheduling and graphical interface for pipeline monitoring, which is a critical aspect of workflow orchestration.

In conclusion, a data engineer's tech stack is a well-curated collection of tools and technologies designed to address various data engineering aspects. Mastery of these elements not only makes you more effective in your role but also increases your marketability to potential employers.

Summary

In this chapter, we have discussed the fundamental elements that comprise the role and responsibilities of a data engineer, as well as the technology stack that supports these functions. From programming languages such as Python and Java to data storage solutions and processing frameworks, the toolkit of a data engineer is diverse and integral to their daily tasks. As you prepare for interviews or take the next steps in your career, a thorough understanding of these elements will not only make you more effective in your role but will also make you more appealing to potential employers.

As we move on to the next chapter, we will focus on an additional crucial aspect of your data engineering journey: portfolio projects. Understanding the theory and mastering the tools are essential, but it is your ability to apply what you've learned in real-world situations that will truly set you apart. In the next chapter, *Must-Have Data Engineering Portfolio Projects*, we'll examine the types of projects that can help you demonstrate your skills, reinforce your understanding, and provide future employers with concrete evidence of your capabilities.

2

Must-Have Data Engineering Portfolio Projects

Getting through a data engineering interview requires more than just knowing the fundamentals. Although having a solid theoretical foundation is important, employers are increasingly seeking candidates who can start working right away. This entails building a portfolio of completed projects that show off the depth and breadth of your abilities in practical settings. In this chapter, we will walk you through the fundamental skill sets that a data engineering portfolio should include and demonstrate, with an example project, where you build an entire data pipeline for a sports analytics scenario.

With a well-designed portfolio, employees can see that you are not just knowledgeable about different concepts but also skilled at putting them to use. By the end of this chapter, you'll have a clear plan for creating projects that stand out from the competition and impress hiring managers and recruiters.

In this chapter, we're going to cover the following topics:

- Must-have skillsets to showcase in your portfolio
- Portfolio data engineering project

Technical requirements

You can find all the code needed for the sports analytics pipeline at https://github.com/PacktPublishing/Cracking-Data-Engineering-Interview-Guide/tree/main/Chapter-2.

Must-have skillsets to showcase in your portfolio

In the rapidly evolving field of data engineering, having a wide and comprehensive skill set is not only advantageous but also essential. As you get ready for your next professional step, you need to make sure your portfolio showcases your abilities in different areas of data engineering.

This section will act as your resource for key competencies that your data engineering portfolio must highlight. There are a lot of different skills you can add to a project, but we will focus on some fundamentals. The following figure shows the different phases of a data pipeline. Each project does not need to have every single element, but your whole portfolio should cover multiple ones:

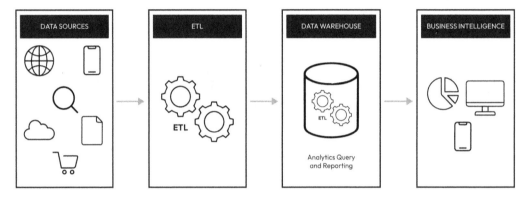

Figure 2.1 – Basic phases of the ETL process

These fundamental abilities demonstrated in your portfolio will make you an attractive candidate to potential employers, regardless of your experience level.

Ability to ingest various data sources

The ability to consistently ingest data from multiple sources is one of the most fundamental tasks in data engineering applications. Data can originate from various platforms and come in a variety of formats. These can include flat files, streaming services, databases, and APIs. Your portfolio needs to show that you can handle this diversity. In this section, we'll look at how to ingest data from various sources, talk about potential problems, and walk you through best practices:

- **Local files:** This includes CSV, Excel spreadsheets, and TXT files. These are files that are normally locally available and are the simplest formats to deal with. However, on the job, you will be most likely dealing with more complex data sources. Websites such as Kaggle, the Google Dataset search engine, data.gov, and the UCI Machine Learning Repository are a few of the various sources for readily available datasets in spreadsheet form.

- **Web page data**: You can use this to build web scrapers that pull data from a web page. For Python users, `BeautifulSoup`, `Selenium`, `Requests`, and `Urllib` are a few libraries you can use to harvest data within HTML. Not all web pages allow for web scraping.

- **Application programming interfaces (APIs)**: APIs allow you to extract live data from applications and websites such as Twitter or `https://www.basketball-reference.com/`. Unlike a web scraper, you can query or select the subsets of data that you would like from an API. These APIs may come with documentation that provides instructions on how to write the code to utilize the API.

- **JavaScript Object Notation (JSON) files**: When extracting data from an API or dealing with nested data in a database, you will encounter JSON files. Be sure you have practiced the ability to handle JSON data.

For any data engineer, ingesting data from multiple sources is an essential skill. Showcasing your proficiency in managing various data sources will make a great impression on potential employers. Being aware of best practices will help you stand out from the competition. These include handling errors, validating data, and being efficient.

Data storage

Once you have ingested data for your project, you should showcase your data storage skills. Whether you're dealing with structured data in relational databases or unstructured data in a data lake, your choice of storage solutions has a significant impact on accessibility, scalability, and performance. Relational databases such as PostgreSQL and MySQL are frequently chosen for structured data because of their **ACID** properties: **Atomicity**, **Consistency**, **Isolation**, and **Durability**. These databases provide the required robustness for transactional systems, enabling complex querying capabilities. In contrast, NoSQL databases such as MongoDB and Cassandra are gaining popularity due to their ability to scale horizontally and accommodate semi-structured or unstructured data, making them ideal for managing large volumes of data that do not neatly fit into tabular structures:

- **Relational SQL databases**: You can store your various structured data sources in a local relational database such as PostgreSQL, MySQL, or SQLite so that it can be queried for later use. Alternatively, you can use cloud databases by using services such as AWS or Azure. You can also create a data model using either the star or transactional method.

The following diagram depicts the star schema:

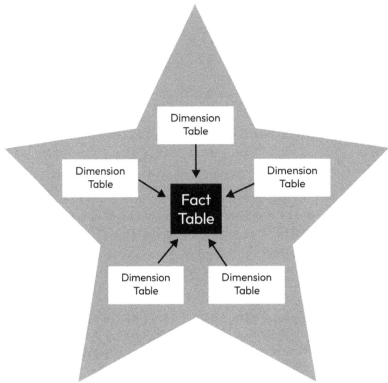

Figure 2.2 – Example visual of the star schema

- **NoSQL databases**: All your unstructured data sources (**Internet of Things (IoT)**, images, emails, and so on) should be stored in NoSQL databases such as MongoDB.

- **Storage architecture**: Practice staging your data in separate zones based on transformation levels:

  - Raw and unprocessed

  - Cleaned and transformed

  - Curated views for dashboarding and reporting

Your portfolio will stand out if you can show that you are capable of managing a variety of data sources, including flat files and APIs. Be sure to highlight certain best practices, including error handling, data validation, and efficiency.

Data processing

Once data has been ingested and stored, the focus shifts to data processing. This is where we transform raw data into a usable form for future analysis. At its core, data processing consists of a series of operations designed to cleanse, transform, and enrich data in preparation for analysis or other business applications. Traditional **Extract, Transform, Load** (**ETL**) procedures are being supplemented or replaced by **Extract, Load, Transform** (**ELT**) procedures, especially when dealing with cloud-based storage solutions.

In data processing, data quality and integrity are also evaluated. Missing values are handled, outliers are examined, and data types are cast appropriately to ensure that the data is reliable and ready for analytics. Stream processing tools such as Kafka and AWS Kinesis are better suited for real-time data flows, enabling immediate analytics and decision-making.

Here are some aspects of the data processing portion that you want to highlight in your projects:

- **Programming skills**: Write clean and reproducible code. You should be comfortable with both object-oriented and functional programming. For Python users, the PEP-8 standard is a great guide.

- **Converting data types**: You should be able to convert your data types as necessary to allow for optimized memory and easier-to-use formats.

- **Handling missing values**: Apply necessary strategies to handle missing data.

- **Removing duplicate values**: Ensure all duplicate values are removed.

- **Error handling and debugging**: To create reproducibility, implement blocks of code to handle anticipated errors and bugs.

- **Joining data**: Combine and merge different data sources.

- **Data validation and quality checks**: Implement blocks of code to ensure processed data matches the source of truth.

Once a data pipeline has been built, you can use a tool such as Apache Airflow to orchestrate and schedule tasks automatically. This will be particularly useful for projects that use datasets that are refreshed periodically (daily, weekly, and so on).

Cloud technology

Since cloud services such as **Amazon Web Services** (**AWS**), Microsoft Azure, and **Google Cloud Platform** (**GCP**) offer flexibility and scalability, they have become essential components of contemporary data engineering. Large data processing and storing can now be accomplished by enterprises at a fraction of the cost and effort opposed to bulky hardware and data centers. The goal of this section is to provide you with an overview of the different cloud-based solutions and how the data engineering ecosystem benefits from them. Practical experience with cloud technologies not only increases your adaptability but also helps you stay up to date with new trends and industry best practices.

Among the top cloud service providers is AWS. The following are some important services for data engineering:

- **S3**: Raw or processed data can be stored using S3, a simple storage service
- **Glue**: An entirely managed ETL solution
- **Redshift**: A solution for data warehousing
- **Kinesis**: Data streaming in real time

GCP provides a range of cloud computing services that are powered by the same internal infrastructure that Google uses for its consumer goods:

- **Cloud Storage**: AWS S3-like object storage solution
- **Dataflow**: Processing of data in batches and streams
- **BigQuery**: A highly scalable, serverless data warehouse

Azure from Microsoft offers a variety of services designed to meet different needs in data engineering:

- **Blob Storage**: Scalable object storage for unstructured data
- **Data Factory**: A service for data integration and ETL
- **Azure SQL Data Warehouse**: A fully managed data warehouse with performance enhancements
- **Event Hubs**: Ingestion of data in real time

In the modern world, cloud technologies are essential for any data engineer to understand. While the features offered by each cloud provider are similar, the subtle differences can have a significant impact on the requirements of your project. Demonstrating your proficiency in navigating and implementing solutions within AWS, GCP, or Azure demonstrates your adaptability to the dynamic field of data engineering to prospective employers.

Portfolio data engineering project

In this section, we will look at an example Azure data engineering project on sports analytics that involves creating a pipeline that ingests, cleans, and visualizes data.

Scenario

You were recently employed by a company (Connect) that has rendered all data-related services to its clients for the past 2 months. You have been attached to a team, but today, you have been given your first job sole project.

A new season of the English Premier League just commenced and your company assigned you to a data engineering job posted by a client. They provided several website links to acquire the data from. Your job is to have all this data extracted, transformed, and loaded into cloud storage and a PostgreSQL database every Saturday and Sunday until the end of the season.

As a data engineer, you should be able to assess the requirements to properly decide which tool to use for each task that would make the process efficient. For example, using Spark to transform 70 rows of data can be counterproductive because that tool is meant for data that contains over a million rows.

The following figure visualizes the project's data pipeline and tools to be used:

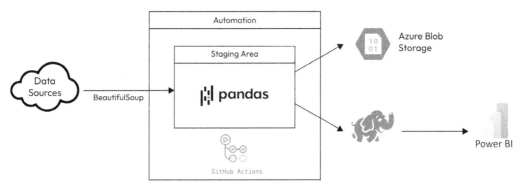

Figure 2.3 – Sports analytics data pipeline

Here's the approach you should take:

1. Go to websites and look at the data to be extracted. Determine which tool you'll use to get the data.
2. Build your web scraping script to get and transform this data.
3. Provision a Blob storage container on Azure and upload your extracted data as a Parquet file.
4. Provision a PostgreSQL database on Azure and upload your extracted data to it.
5. Automate this process and schedule it with GitHub Actions to run every Saturday and Sunday until the end of the season.
6. Connect a Power BI service to your PostgreSQL database and do some **Exploratory Data Analysis (EDA)**.

Let's look at the steps for this project.

Step 1: Examine the data sources.

Go to the following sites:

- League table: `https://www.bbc.com/sport/football/premier-league/table`
- Top scorers: `https://www.bbc.com/sport/football/premier-league/top-scorers`
- Detailed top scorers: `https://www.worldfootball.net/goalgetter/eng-premier-league-2023-2024/`
- Player table: `https://www.worldfootball.net/players_list/eng-premier-league-2023-2024/`
- All time table: `https://www.worldfootball.net/alltime_table/eng-premier-league/pl-only/`
- All-time winner (clubs): `https://www.worldfootball.net/winner/eng-premier-league/`
- Top scorers per season: `https://www.worldfootball.net/top_scorer/eng-premier-league/`
- Goals per season: `https://www.worldfootball.net/stats/eng-premier-league/1/`

After inspecting the sites, we can see that they are in table format and they are all static sites as opposed to dynamic sites. The key difference between both is that static websites have stable content, where every user sees the same thing on each page, such as a privacy policy, whereas dynamic websites pull content on the fly, allowing its content to change with the user. If it was a dynamic site, selenium would be more suitable.

Step 2: Build your web scraping script to ingest and transform your data.

Follow these steps:

1. We'll build our scraper with Python functions in a file called `scrape.py`. This way, we can call and run the function from another script just by importing it. But first, let's look at some basic code we can use to scrape table content from most static sites using Beautiful Soup with little to no cleaning:

   ```
   import requests
   from bs4 import BeautifulSoup
   import pandas as pd
   ```

```
url = 'https://www.bbc.com/sport/football/premier-league/table'
#the page link we want to extract from
headers = []
#a list that stores our header names
page = requests.get(url)
#makes a request to the webpage and returns the html content
soup = BeautifulSoup(page.text, "html.parser")
#we can use this to clean and sort through the html content to
get what we need
table= soup.find("table", class_="ssrcss-14j0ip6-Table
e3bga5w5")
#find the contents in this part of the html content with that
class_name
#we use _class because the class is a keyword in python
for i in table.find_all('th'):
#finds all the html tags th which holds the header details
 title = i.text
#gets the text content and appends it to the headers list
 headers.append(title)
league_table = pd.DataFrame(columns = headers)
#creates a dataframe with the headers
for j in table.find_all('tr')[1:]:
#finds all the content with tr tag in the table
 row_data = j.find_all('td')
 row = [i.text for i in row_data]
 length = len(league_table)
 league_table.loc[length] = row
#gets them by row and saves to the league_table dataframe
print(league_table)
```

The following screenshot shows one of the tables that will be extracted:

Premier League Table

Q Enter a team or competition

Position	Team	Played	Won	Drawn	Lost	For	Against	GD	Points	Form
1	Tottenham Hotspur	9	7	2	0	20	8	12	23	W W D W W W
2	Manchester City	9	7	0	2	19	7	12	21	W W W L L W
3	Arsenal	9	6	3	0	18	8	10	21	W W D W W D
4	Liverpool	9	6	2	1	20	9	11	20	W W W L D W
5	Aston Villa	9	6	1	2	23	13	10	19	L W W W D W
6	Newcastle United	9	5	1	3	24	9	15	16	L W W W D W
7	Brighton & Hove Albion	9	5	1	3	22	18	4	16	W W W L D L
8	Manchester United	9	5	0	4	11	13	-2	15	L L W L W W
9	West Ham United	9	4	2	3	16	16	0	14	W L L W D L
10	Chelsea	9	3	3	3	13	9	4	12	L D L W W D
11	Crystal Palace	9	3	3	3	7	11	-4	12	W L D W D L

Figure 2.4 – Premier League table of team statistics

2. To get the content you need, right-click anywhere on the site and select **inspect**. This will open a developer console for you to see the HTML content of the page. It would look something like the following visual:

Figure 2.5 – What you will see in the page inspection area

As you can see, the `<th>` tag is under the `tr` tag, which is under the `thead` tag, which holds all the header names. If you sort further, you'll see the `tbody` tag, which holds all the row content. As you can see, they are all under the table content with the `ssrcss-1y596zx-HeadingRow e3bga5w4'` class name.

The output of the preceding code will give us the output shown in the following screenshot:

```
In [1]: import requests
        from bs4 import BeautifulSoup
        import pandas as pd

        url = 'https://www.bbc.com/sport/football/premier-league/table'
        #the page link we want to extract from
        headers = []
        #a list that stores our header names
        page = requests.get(url)
        #makes a request to the webpage and returns the html content
        soup = BeautifulSoup(page.text, "html.parser")
        #we can use this to clean and sort through the html content to get what we need
        table= soup.find("table", class_="ssrcss-14j0ip6-Table e3bga5w5")
        #find the contents in this part of the html content with that class_name
        #we use _class because the class is a keyword in python
        for i in table.find_all('th'):
        #finds all the html tags th which holds the header details
          title = i.text
        #gets the text content and appends it to the headers list
          headers.append(title)
        league_table = pd.DataFrame(columns = headers)
        #creates a dataframe with the headers
        for j in table.find_all('tr')[1:]:
        #finds all the content with tr tag in the table
          row_data = j.find_all('td')
          row = [i.text for i in row_data]
          length = len(league_table)
          league_table.loc[length] = row
        #gets them by row and saves to the league_table dataframe
        print(league_table)
```

	Position	Team	Played	Won	Drawn	Lost	Goals For \
0	1	Tottenham Hotspur	8	6	2	0	18
1	2	Arsenal	8	6	2	0	16
2	3	Manchester City	8	6	0	2	17
3	4	Liverpool	8	5	2	1	18
4	5	Aston Villa	8	5	1	2	19
5	6	Brighton & Hove Albion	8	5	1	2	21
6	7	West Ham United	8	4	2	2	15
7	8	Newcastle United	8	4	1	3	20
8	9	Crystal Palace	8	3	3	2	7
9	10	Manchester United	8	4	0	4	9
10	11	Chelsea	8	3	2	3	11
11	12	Fulham	8	3	2	3	8
12	13	Nottingham Forest	8	2	3	3	8
13	14	Wolverhampton Wanderers	8	2	2	4	9
14	15	Brentford	8	1	4	3	11
15	16	Everton	8	2	1	5	9
16	17	Luton Town	8	1	1	6	6
17	18	Burnley	8	1	1	6	7
18	19	AFC Bournemouth	8	0	3	5	5
19	20	Sheffield United	8	0	1	7	6

Figure 2.6 – Screenshot of the initial web scraping results

3. Now, we can build our `scrape.py` file, which contains all the functions we need to extract the data the client asked for. Create a Python file called `scrape.py`. This can be found in this book's GitHub repository at `https://github.com/PacktPublishing/Cracking-Data-Engineering-Interview-Guide/blob/main/Chapter-2/scrape.py`.

Step 3: Provision a Blob storage container on Azure and upload your extracted data to it as a Parquet file.

With the DataFrame that each function in the `scrape.py` returns, we can save it to Blob storage as a Parquet file as opposed to saving it to our local system.

Parquet is a popular choice for storing and processing large datasets, but it's not the only option available. Other file formats, such as **Optimized Row Columnar** (**ORC**), Avro, and JSON, also have their advantages and use cases. Here's why you might choose Parquet over these other file formats:

- **Parquet versus ORC**: Both Parquet and ORC are columnar storage formats designed for similar purposes, such as big data analytics. They offer similar benefits in terms of compression, predicate pushdown, and schema evolution.

  The choice between Parquet and ORC often depends on the specific ecosystem you're working in. For example, Parquet might be preferred if you're using tools such as Apache Spark, while ORC might be better suited for environments such as the Hadoop ecosystem with Hive and Tez.

- **Parquet versus Avro**: Avro is a row-based data serialization format that focuses on data interchange and is used in various tools and frameworks, including Apache Kafka and Apache Hadoop:

  - Parquet's columnar storage provides better compression and query performance advantages, especially for analytical workloads that involve aggregations and filtering

  - Avro's simplicity and support for schema evolution make it suitable for scenarios where you need a more lightweight format and don't require the same level of query performance optimization that Parquet offers

- **Parquet versus JSON**: JSON is a human-readable data format and is widely used for data interchange. However, JSON is not as efficient for storage and processing as columnar formats such as Parquet:

  - Parquet's columnar storage and advanced compression techniques make it much more space-efficient and better suited for analytical workloads that involve reading specific columns

  - JSON might be preferred in cases where human readability and ease of use are more important than storage and processing efficiency

- **Parquet versus CSV**: Parquet files are smaller than CSV files, and they can be read and written much faster. Parquet files also support nested data structures, which makes them ideal for storing complex data.

CSV is a row-based data format that is simple to read and write

1. Now, to create our program, which pushes this DataFrame to our created container as a Parquet file, I named my container `testtech` and did the following:

    I.   First, I imported the necessary modules and classes:

```
from scrape import *
#imports our functions from scrape.py so we can call the
functions from here
import pandas as pd
import pyarrow as pa
import pyarrow.parquet as pq
from io import BytesIO
from azure.storage.blob import BlobServiceClient, BlobClient,
ContainerClient
```

    II.   Then, I loaded the environment variables from a `.env` file:

```
# List of functions to process and upload
functions = [league_table, top_scorers, detail_top, player_
table, all_time_table, all_time_winner_club, top_scorers_
seasons, goals_per_season]

# Function to upload data to Azure Blob Storage
def to_blob(func):
    """
    Converts the output of a given function to Parquet format and
    uploads it to Azure Blob Storage.

    Args:
    func (function): The function that retrieves data to be
    processed and uploaded.

    Returns:
    None

    This function takes a provided function, calls it to obtain
    data, and then converts the data into
    an Arrow Table. The Arrow Table is serialized into Parquet
    format and uploaded to an Azure Blob
    Storage container specified in the function. The function's
    name is used as the blob name.

    Example:
    Consider the function "top_scorers". Calling "to_blob(top_
```

```
scorers)" will process the output
of "top_scorers", convert it to Parquet format, and upload it
to Azure Blob Storage.
"""

# Getting the name of the function
file_name = func.__name__
# Calling the function to retrieve data
func = func()

# Converting DataFrame to Arrow Table
table = pa.Table.from_pandas(func)

# Creating a buffer to store Parquet data
parquet_buffer = BytesIO()
# Writing the Arrow Table data to the buffer in Parquet format
pq.write_table(table, parquet_buffer)

# Retrieving Azure Blob Storage connection string
connection_string ='your container connection string'
# Creating a Blob Service Client
blob_service_client = BlobServiceClient.from_connection_
string(connection_string)

# Specifying container name and blob name
container_name = "testtech"
blob_name = f"{file_name}.parquet"
# Creating a Container Client
container_client = blob_service_client.get_container_
client(container_name)
# Creating a Blob Client
blob_client = container_client.get_blob_client(blob_name)

# Uploading Parquet data to Azure Blob Storage
blob_client.upload_blob(parquet_buffer.getvalue(),
    overwrite=True)
# Printing a success message
print(f"{blob_name} successfully updated")

# Loop through the list of functions and upload data
for items in functions:
    to_blob(items)
```

2. To get your container connection string, scroll down on the provisioned resource page and copy your string from under **Access key**. After the code runs successfully, check your storage container. You should see something similar to the following:

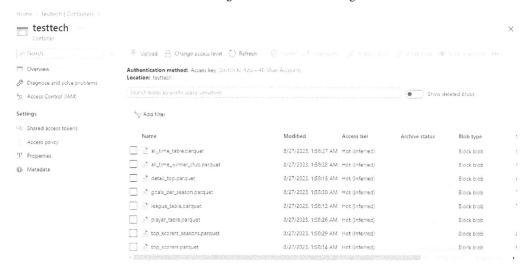

Figure 2.7 – Storage container with added data

It is advisable to use a cool or cold access tier as it saves more money.

Step 4: Provision a PostgreSQL database server on Azure and upload your extracted data to it.

Provisioning a PostgreSQL database server on Azure can be done by following these steps:

1. Search for `Azure database for Postgres flexible server` in the search bar at the top of the Azure portal, then click **Create the resource**.

2. On the **Basics** tab, choose your Azure subscription if it hasn't already been indicated for you. Then, go ahead and choose your resource group or create one if there are none.

3. Enter your server name. Your workload should be on development unless you want to make the project bigger. Type in an admin name of your choice and your password. Confirm this and press *Enter*.

4. On the **Networking** page, leave everything else as-is except for **Firewall Rules**. Here, you can either allow connection by adding your IP address or add someone else's to allow access from all IP addresses, which is not safe. Then, click **Review + Create**.

To push our football DataFrames directly to a database, we need to implement some code:

I. Import the necessary modules and functions:

```
from scrape import *
import pandas as pd
from sqlalchemy import create_engine
import psycopg2
```

II. List the functions to process and push them to the database:

```
functions = [league_table, top_scorers, detail_top, player_
table, all_time_table, all_time_winner_club, top_scorers_
seasons, goals_per_season]
```

III. Retrieve the database connection string from environment variables:

```
conn_string = 'postgres://user:password@hostname:5432/database-
name'
# database-name is sometimes "postgres" incase you get into an
error
# Creating a database engine
db = create_engine(conn_string)
# Establishing a database connection
conn = db.connect()
```

IV. Loop through the list of functions and push data to the database for fun in functions:

```
# Get the name of the current function
 function_name = fun.__name__

# Call the function to get the DataFrame
 result_df = fun()
```

5. Push the DataFrame to the database table with the function name as the table name:

```
result_df.to_sql(function_name, con=conn, if_exists='replace',
index=False)

# Print a message indicating data has been pushed for the
current function
 print(f'Pushed data for {function_name}')

# Close the database connection
conn.close()
```

You can get your connection string from your **Connect** tab and fix it into the format shown previously.

We need one more Python script, called `main.py`, to run `push_to_blob.py` and `push_to_database.py` instead of running them separately:

```
# Import the os module for operating system-related
functionality
import os

# Run the 'push_to_blob.py' script using the system shell
os.system('python push_to_blob.py')

# Run the 'push_to_database.py' script using the system shell
os.system('python push_to_database.py')
```

Step 5: Automate this process and schedule it with GitHub Actions to run every Saturday and Sunday until the end of the season.

Follow these steps:

1. Push all your working code to a GitHub repository.
2. Go to **Settings** in your GitHub repository, scroll down to **Secrets and Variables**, and click **Actions**.

   Now, we will work on our YAML file, which will schedule and automate the extraction, transformation, and loading of this data.

   To use our hidden environment variables inside our code, we need to use the `dotenv` library. GitHub serves as the `.env` file:

```
from dotenv import load_dotenv
import os

load_dotenv()
# for your push_to_database.py
conn_string = os.getenv('CONN_STRING')

#for your push_to_blob.py

connection_string = os.getenv('BLOB_CONN_STRING')
# Name of the GitHub Actions workflow
name: update league data

# Define when the workflow should run
on:
  schedule:
```

```
- cron: '0 0 * * 6,0' # Runs at 00:00 every Saturday and Sunday
 workflow_dispatch:

# Define the jobs to be executed within the workflow
jobs:
 build:
 # Specify the runner environment
 runs-on: ubuntu-latest
 steps:

 # Step 1: Checkout the repository content
 - name: checkout repo content
 uses: actions/checkout@v2 # Checkout the repository content to
GitHub runner

 # Step 2: Setup Python
 - name: setup python
 uses: actions/setup-python@v4
 with:
 python-version: '3.9' # Install the specified Python version

 # Step 3: Install Python packages
 - name: install python packages
 run: |
 python -m pip install --upgrade pip
 pip install -r requirements.txt

# Step 4: Execute extract script
 - name: execute extract script
 env:
 BLOB_CONN_STRING: ${{ secrets.BLOB_CONN_STRING }}
 # Set environment variable from GitHub secrets
 CONN_STRING: ${{ secrets.CONN_STRING }}
 # Set environment variable from GitHub secrets
 run: python main.py # Run the 'main.py' script
```

Step 6: Connect a Power BI service to your PostgreSQL database and do some exploratory data analysis.

This part will be completely up to you. Remember that this part is not necessary to the client as they already have a data analyst waiting to get the insights.

You can use any dashboarding tool you like, so long as it can connect to a cloud database.

Summary

In this chapter, we covered the fundamentals of data engineering that aspiring engineers can include in their portfolios. We covered the general phases of building data pipelines, including data ingestion, processing, cloud technologies, and storage. We concluded with a sports analytics example project to give you a practical way to put these skills to use. By mastering and showcasing these essential skill sets via portfolio projects, you will not only stand out during interviews, but you will also establish a solid foundation for your data engineering career.

Upon moving on to our upcoming chapter, *Building Your Data Engineering Brand on LinkedIn*, we will concentrate on converting these technical achievements into an engaging LinkedIn profile. Having a strong online presence can help you attract recruiters and hiring managers.

Building Your Data Engineering Brand on LinkedIn

According to HubSpot, approximately 85% of jobs today are filled through networking. And one of the best online platforms for professional networking is LinkedIn. Now, in the remote working era, platforms such as LinkedIn allow you to connect with valuable people in the data engineering field such as managers, current data engineers, recruiters, and executives. Building rapport and relationships with the aforementioned groups allows you to attract opportunities. Companies need to hire diverse data engineering talent. And when you don't have any initial connections to the data field, there's no better way to build your network and facilitate job offers than by building a brand on social media.

While this chapter will focus on LinkedIn, these tips can be applied to other social networking platforms such as Twitter/X. For the sake of simplicity, we will only focus on LinkedIn.

We will cover the following main topics:

- Optimizing your LinkedIn profile
- Crafting your About Me section
- Developing your brand

Optimizing your LinkedIn profile

Having a well-developed LinkedIn profile can lead to increased visibility to hiring managers and recruiters looking for talent. If recruiters and HR professionals use LinkedIn as their major resource for locating talent, this can increase your visibility. Using LinkedIn to share your skills and ideas with others in your field will also help you stand out as an industry expert and expand your professional network.

In this section, we will examine the profile header, the banner, the **About Me** section, and the profile photo in greater detail. We'll review strategies for making the most of these sections and adapting them to your data engineering career goals. You will learn all you need to know to create a stellar

LinkedIn profile that will help you stand out as a data engineer, get seen by recruiters, and boost your personal brand.

Your profile header consists of the following main components:

- Profile picture
- Banner
- Headline
- Most recent employment and education

The following screenshot displays an example of a well-designed LinkedIn header with a customized banner, appropriate headshot, and headline. Notice all elements are meant to highlight your talents and abilities:

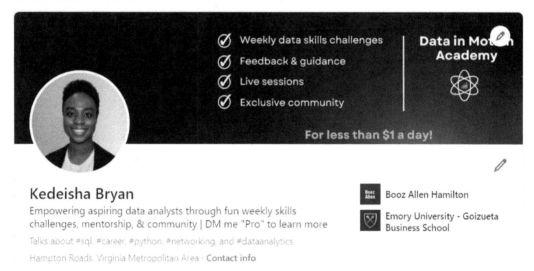

Figure 3.1 – Example of a well-presented LinkedIn header

Many overlook this section as many profile headers often look similar. But, making simple edits to your header can help it stand out whenever potential employers check your profile.

Your profile picture

Your profile photo plays a crucial role in shaping how others perceive you at first glance. According to a study, LinkedIn profiles with pictures attract more than twice as many views and three times as many messages versus those without.

A data engineer's profile photo should convey competence and friendliness. Here's how to make sure your photo makes a good impression:

- **Quality**: First and foremost, the quality of your image is what matters most. Avoid poor image quality, such as blurring or pixels.

- **Professional attire**: You want to present yourself in a professional manner. While a suit is not required, you should dress in a business casual manner at the very least.

- **Simple background**: Keep the background simple so that people's eyes are drawn to your face. It's better to have a plain or neutral backdrop. Distractingly cluttered backdrops should be avoided.

- **Close-up and well-lit**: Your photo should be a close-up of your face, shot from the top of your shoulders to above your head, and lit appropriately.

- **Friendly expression**: Display a natural, welcoming expression. Being friendly and approachable is essential. Sometimes, all that's needed is a friendly grin.

Once you have a photo that you are satisfied with, you can follow these steps to remove your background:

1. The first step is to remove the background. If you do not already have a Canva Pro account, head over to any photo background removal site, such as `remove.bg` or `https://www.photoroom.com/tools/background-remover`. There, you will upload your photo. You should end up with a photo with just yourself with a checkered background signaling that the background was removed.

2. Afterward, head over to Canva and create a new design of 1,600 x 1,600 pixels (or another if you'd like).

3. Once there, upload your new photo with the background removed. On the left-hand side, click **Elements**.

4. There, you will search for the term `backgrounds`.

5. Now, to implement a gradient background, search for `gradient background`.

6. Once you have your preferred background, drag it to the canvas and allow it to fill the entire space. Then, move it to the background by clicking the position at the top. There, you can arrange the order and place your headshot on the top layer.

The following screenshot depicts how to arrange the order of layers in your photo:

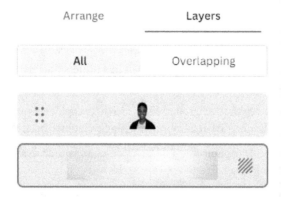

Figure 3.2 – The order in which to arrange the layers of your photo

Remember that your LinkedIn photo should do more than just display your physical appearance; it should also reflect your personality. Before you even speak with a possible employer or connection, an image that is professional, clear, and friendly can go a long way toward establishing trust and rapport. It's a chance to make a good impression. Spending the time and energy to get it perfect will greatly increase the usefulness of your LinkedIn profile.

Your banner

A banner, often called a backdrop photo, is an enormous image that appears at the very top of your profile. It's excellent real estate that can boost your personal brand, yet it's underutilized most of the time. It's possible that a data engineer would choose a picture that has something to do with their profession. You can add pictures of some certifications you may have such as AWS or Azure. If you're good with design, you might make a unique graphic with your details and a catchy slogan or prompt already on it. Make sure the picture you use is of professional quality. You want your banner to draw attention to your title and make it stand out.

It can be easy to overlook your profile banner. The worst thing you can have on your profile is an empty one, as in the following screenshot:

Figure 3.3 – An example of a profile with an empty banner

Your banner should be personalized to visualize who you are and your skills. While LinkedIn has several to choose from, I highly recommend creating your own using a free Canva account.

On the Canva site, click on **Create a design**. In the drop-down menu, select **LinkedIn Background Photo**, as seen in the following screenshot:

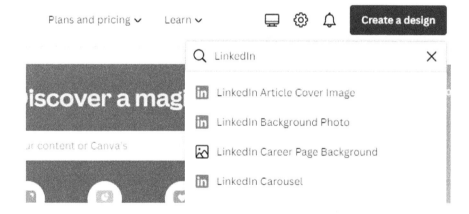

Figure 3.4 – Selecting Create a design to initiate the process of creating a custom LinkedIn banner

Once the canvas uploads, you have all the pre-created banner templates to your left. You are free to customize your banner as you wish. Once you have one that you are satisfied with, upload your new banner to your LinkedIn profile.

If you take the time to customize your banner, you'll be able to make better use of the available visual real estate and give your profile a more personal feel. The key is to make an impression that stands out and is consistent with your data engineer brand. Remember that the banner is one of your profile's many storytelling opportunities.

Header

The default headline is your work title and employer, but you can change this if you'd like. It doesn't have to be limited to your official title. It's like a condensed version of your elevator pitch for potential employers. The most effective headlines manage to be both brief and informative. They define your current position, highlight your abilities, and highlight relevant experience. You can include industry-specific keywords such as *Data Engineering*, *Big Data*, and *ETL* but don't only compile a collection of cliches. Write in a narrative fashion that organically incorporates these terms. One possible headline is *Data engineer using big data tools to uncover actionable business insights*. This highlights your contribution, area of expertise, and role in the organization; we can see example details in the following screenshot:

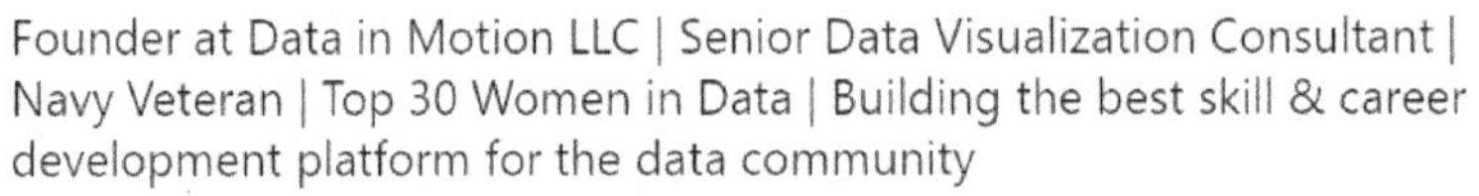

Founder at Data in Motion LLC | Senior Data Visualization Consultant | Navy Veteran | Top 30 Women in Data | Building the best skill & career development platform for the data community

Figure 3.5 – Example of a well-defined header

One big mistake people make with their headlines is attaching the word *aspiring* to any data profession they are applying to. You've been doing the projects and creating your own experiences. Avoid *aspiring* at all costs.

Here are some examples for aspiring data engineers with no professional engineering experience:

- Data Engineer | Proficient in Database Design and Management | SQL | Python | Microsoft Azure

- Data Engineer | Skilled in Data Processing and Big Data Technologies | SQL | Python | AWS | Hadoop | Spark

- Data Engineer | Experienced in Data Integration and Analysis | SQL | Python | AWS | Kafka

- Data Engineer | Cloud Computing and Distributed Systems

LinkedIn also offers the ability to add a custom link in your header to direct viewers of your profile to another site:

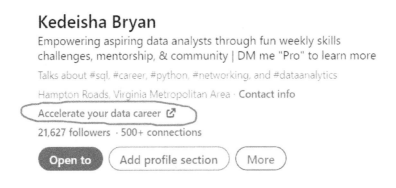

Figure 3.6 – Custom link in your profile header

With the options in the preceding screenshot we can do the following:

- Under the profile header settings, you can add a link to your portfolio to direct prospective hiring managers and recruiters to your work

- Add a link to your portfolio under the **Custom Action** section of the profile header settings

- Under **Custom Action** in the settings, you can place the URL and a custom name of the link

These are just a few examples of LinkedIn headlines but allow yourself to be creative. In the next section, we will discuss how to optimize your **About Me** section.

Crafting your About Me section

Your LinkedIn **About Me** section should serve as a professional autobiography. You can give prospective employers a glimpse into who you are as a person as well as a data engineer by detailing your professional background and accomplishments here.

In the field of data engineering, where many different profiles may appear to have comparable talents and expertise, this part becomes increasingly important. It's a chance to highlight your individuality and demonstrate why you're not like every other data engineer out there. This is your opportunity to tell a story about yourself that highlights your technical abilities and accomplishments.

In the next section, we'll go through some tips for writing an effective **About Me** section. We'll talk about how to present your data engineering credentials in a way that stands out from the crowd and makes you an attractive candidate for a job.

Initial writing exercise

First, think of three to five things about yourself that you would like any hiring manager to know about you after an interview. Avoid only mentioning three technical tools. Think about more unique soft skills, such as the following:

- Mentoring and teaching

- Leading teams

- Problem-solving

- Project management

- Adaptability

Once you think about the instances in your professional life where you've demonstrated your top three to five capabilities, be sure to highlight them in your **About Me** section. The following are some guidelines for writing an eye-catching **About Me** section:

- **Opening statement**: Start with a strong opening statement summarizing your identity as a data engineer. What you do now, your most important abilities, and what you bring to an organization are examples. For example, you could write something such as *A results-oriented data engineer capable of turning raw data into meaningful insights that fuel company expansion*.

- **Share your experience**: Expand on your introductory remarks by discussing your professional background in further detail. When did you first get curious about data engineering? Just how did you get to where you are today? Can you describe the fresh ideas you bring to the table?

- **Highlight key skills and achievements**: In this section, you can highlight the talents and accomplishments that have helped you succeed. Whenever possible, provide numbers to back up your claims of success. For instance, *I built a data processing system that increased data accuracy by 30 percent, which led to more trustworthy insights about my company*.

- **Put your skills on display**: List the software, databases, and programming languages you are comfortable working with. You can provide context for your knowledge by discussing projects or jobs in which you employed these abilities.

- **Aims and desires**: Describe your intentions and objectives in the workplace. In what ways do you hope to advance your career? What motivates you most in a project? Potential employers can see here whether their goals and yours are congruent.

- **Personal touch**: Adding a little of your personality is always appreciated. You may want to explain how your fascination with solving mysteries led you to work with data, or how your insatiable appetite for knowledge has kept you abreast of the rapid changes in data engineering. This adds a human touch to your profile, which increases its recall value.

- **Call to action**: Encourage people to interact with you, view your portfolio, and contact you about potential opportunities.

The **About Me** section should be written in the first person and in a casual tone. A professional overview should sound different from a promotional pamphlet. The purpose is to help establish a unique professional brand identity for the reader while also highlighting your professional experience.

Now that you understand how to create your **About Me** section, we will dive into strategies to further develop your professional brand with content.

Developing your brand

Your personal brand is more than just a list of your past employment and skill sets. It's how you want the world to perceive you, and it's made up of your individual experiences, values, and objectives. In the highly competitive field, a well-established personal brand can be the determining factor in landing a position. LinkedIn is a must-have resource for any professional serious about building their brand. It gives you many opportunities to network with other professionals and share your unique brand of professionalism. Branding yourself on LinkedIn goes beyond merely having a well-written profile; it also entails sharing your expertise, commenting on relevant posts, and carefully adding connections.

In the next section, we'll explore several methods for enhancing your brand. You will discover the best practices for maximizing your visibility, from creating content to participating in relevant discussions.

Posting content

In addition to having a well-written profile, being active is essential to building a name for yourself there. You may show off your knowledge, interact with your network, and keep yourself in the minds of your contacts (including possible employers) by regularly sharing content. The following are tips you can use to start posting on LinkedIn:

- **Posting regularly**: You can post about your portfolio projects and topics you are learning about, or detail your current journey in landing your next data engineering position.

- **Writing articles**: If you're ready for the challenge, producing in-depth articles on LinkedIn can significantly increase your exposure and authority. These may be in-depth analyses of particular data engineering topics, analyses of specific projects, lessons learned, or even tips for those seeking to build a data engineering profession. Sharing such materials establishes you as an authority in your field and increases your exposure on social media and search engines.

- **Showcasing projects**: If you have completed an important project, whether for school or a personal project, by all means, share it with the world. In a blog post or article, you may document the project's goals, methodology, technologies, obstacles encountered, and final results. If you have a GitHub repository, you can also provide a link to that for people interested in learning more. This shows prospective employers that you can deliver results and gives them insight into your thought process.

- **Community engagement**: Broadcasting your opinions is only part of sharing content. Interacting with other people is also essential. Take the time to interact with the people who take the time to comment on your postings, and vice versa. This kind of back-and-forth raises your profile and facilitates connections with like-minded experts.

- **Consistency**: Maintaining uniformity is crucial while establishing your credibility in your field. Aim for once a week, at a minimum, to post something online. Keeping a consistent presence requires careful planning, which you can do with a content schedule.

Sharing content and projects is not only about showing off your knowledge; it should also benefit your network, spark conversations, and help you learn more about your area. Each piece of content you publish or complete contributes to your professional identity. Sharing carefully and frequently will help you become known as an expert, leader, and team player in data engineering. This will make you stand out to potential employers and earn you respect from your peers in the workplace.

Building your network

Personal branding relies heavily on developing a LinkedIn network. It's not about the *quantity* of your connections but the *quality* of contacts in your field; you also need to build strong relationships with those contacts so that you can receive help, advice, and new opportunities. Your data engineering network can consist of people like you: engineers, industry leaders, recruiters, data scientists, analysts, and so on.

The following are tips you can use to start networking more effectively using LinkedIn:

- **Connecting with professionals**: If you want to network with other professionals in your sector or whose work you respect, don't be shy about sending them connection requests. You should make an effort to customize your requests. Explain who you are and why you want to get in touch briefly. This demonstrates that you read their profile carefully and care about them. Provide a friendly and personalized message.

- **Alumni connections**: Maintaining contact with fellow graduates from your institution can be helpful. Conversation starters can be easier if you and your new contacts have something in common to talk about. They may have useful information to share about how to go forward in your chosen field, current trends, or upcoming opportunities.

- **Participating in groups**: Groups on LinkedIn offer a great way to network with other professionals in the data engineering industry and adjacent industries. Also, joining private data communities outside of LinkedIn allows you to network with other professionals in your field, share your expertise, and gain insight from the experiences of others. Getting your name out there among working professionals is another big benefit.

- **Engaging with content**: Submit more than a simple *like* when someone posts something you find interesting. This not only reveals your dedication to the subject matter but it also has the potential to start a dialogue that could lead to a fruitful new relationship.

- **Maintaining contact**: Expanding your network is an ongoing process. It needs constant care and attention. Share your thoughts on your connections' status updates, offer your congratulations on promotions or other accomplishments, and don't be afraid to reach out with a helpful article or a simple "How are you doing?" This is great for keeping in touch with colleagues and expanding your network.

- **Strategic expansion**: Networking with others in your area is essential, but you should also look at expanding into data science, machine learning, and business intelligence. This interdisciplinary group can broaden your perspective on the business world and the careers available to you.

After discussing the importance of creating a stellar profile, publishing useful material, and expanding your network, we will go on to one of the most proactive LinkedIn strategies: cold messaging. Despite the impersonal moniker, cold messaging is a potent technique for making connections and finding success.

We'll go into the science and art of cold messaging in the next part, discussing topics such as target selection, message creation, and following up. Learning how to send successful cold messages can greatly speed up your career journey in data engineering, whether you're looking for advice, an opportunity, or just to expand your network.

Sending cold messages

Sending cold messages is a delicate art. It can come off as spammy or disrespectful if not handled properly. However, if done correctly, it can lead to previously inaccessible interactions and opportunities. Here's how to get the most out of your cold messages:

- **Identify the right people**: Before you start writing, be sure you're addressing your message to the correct people. This may be a senior data engineer whose path you like, a possible employer, or a colleague whose work you find compelling. Take advantage of LinkedIn's *search* and *filter* tools to zero in on the people whose professional or personal interests align with your own.

- **Do your research**: Before sending a cold message, do some digging into the recipient's profile. Find out what they do and how they got there, what they've worked on and posted, and so on. Having this information at your disposal will allow you to tailor your message and show your enthusiasm.

- **Craft your message**: When formulating your message, keep it simple and direct. Introduce yourself and the reason for contacting them pleasantly. Add some personality to your message by highlighting a specific aspect of the recipient's profession or career that has inspired you or interested you. Make it clear what you want to get out of the meeting, whether it's guidance, insights, or the chance to talk about opportunities.

- **Call to action**: Don't leave things hanging at the end; instead, provide a call to action. Inquire as to whether or not they have time for a quick phone chat, LinkedIn message exchange, or email discussion. It's easier to get a *yes* when you give them concrete choices.

- **Follow-up**: Do not lose hope if you do not receive a response to your follow-up. It's easy for people to miss your message if they're too busy. After a week or so, it's appropriate to send a courteous follow-up message. If they continue to ignore you, though, it's time to move on.

Sending a cold message on LinkedIn might be a great approach to taking charge of your career. Connecting with others in the data engineering community can help you grow your professional network, obtain insightful feedback on your work, and open doors to new possibilities if you take the time to do so. It's a preventative measure that, if carried out correctly, can work wonders for your professional standing.

Summary

In this chapter, we've covered the many steps involved in developing your data engineer LinkedIn profile. We began by discussing how to get the most out of your LinkedIn profile by emphasizing the value of a catchy headline, a polished banner, a thoughtful profile image, and an interesting **About Me** section. Then, we dove into the process of disseminating your expertise via content. We talked about how posting on a frequent basis, writing in-depth articles, and showcasing projects may not only show off your knowledge and skills but also keep you at the front of the minds of your contacts.

Next, we discussed the vital topic of constructing your network, stressing the significance of making contact with the proper professionals, joining the correct communities, actively consuming relevant information, and keeping your relationships alive through consistent communication. Finally, we talked about how to properly use cold messages. Finding and researching possible contacts, writing tailored messages, maintaining your professionalism, and following up effectively were all topics we covered.

Each of these tactics is designed to bolster your LinkedIn profile and help you stand out in the competitive profession of data engineering. Implementing these methods can help you build a strong online presence, which in turn will lead to a plethora of options for advancing your data engineering career.

In the next chapter, we will discuss how to prepare for your behavioral interviews.

4

Preparing for Behavioral Interviews

Employers increasingly use behavioral interviews to evaluate candidates' suitability for a specific position. Unlike technical interviews, which consider a candidate's technical knowledge, behavioral interviews assess interpersonal, problem-solving, and decision-making abilities.

This chapter will delve deeply into behavioral interviews and examine the six categories of commonly posed behavioral questions. We will also discuss effectively responding to these queries using the STARR method, a tried-and-true technique for organizing precise and concise responses.

In addition, we will discuss the significance of measuring cultural compatibility and how to prepare for this aspect of the interviewing process. Finally, we will provide examples of frequently posed behavioral interview questions and the most effective responses.

By the conclusion of this chapter, you will have a firm grasp of how to approach behavioral interviews, what to expect, and how to answer questions that will set you apart from other candidates.

We will cover the following topics in this chapter:

- Identifying six main types of behavioral questions to expect
- Assessing cultural fit during an interview
- Utilizing the STARR method when answering questions
- Reviewing the most asked interview questions

Identifying six main types of behavioral questions to expect

Behavioral interview questions are designed to elicit information about how a candidate has handled specific situations to predict how they may perform in similar cases in the future. Employers commonly ask six main types of behavioral interview questions:

- **Situational**: These questions present hypothetical scenarios and ask the candidate how they would respond—for example, "How would you handle a difficult customer who is demanding a refund for a product they damaged themselves?".

- **Behavioral**: These questions ask the candidate to describe a specific situation they faced in the past and how they handled it—for example, "Tell me about a time when you had to deal with a difficult coworker."

- **Goal-oriented**: These questions ask the candidate to describe a specific goal they set for themselves and how they achieved it—for example, "Tell me about a time when you set a challenging goal for yourself and what steps you took to achieve it."

- **Teamwork**: These questions ask the candidate to describe a situation where they worked effectively as part of a team—for example, "Tell me about a time when you collaborated with a team to achieve a common goal."

- **Leadership**: These questions ask the candidate to describe a situation where they demonstrated leadership skills—for example, "Tell me about a time when you had to lead a team through a challenging project."

- **Conflict resolution**: These questions ask the candidate to describe a situation where they had to resolve a conflict with another person or group—for example, "Tell me about a time when you had to mediate a conflict between two team members."

By understanding the six types of behavioral questions, you can better prepare for your interview and be ready to provide clear, concise, and relevant answers that showcase your skills and abilities. The following section will discuss measuring cultural fit during behavioral interviews.

Assessing cultural fit during an interview

As organizations attempt to create diverse and inclusive work cultures that line with their values and mission, measuring for cultural fit has become an increasingly significant component of the recruiting process. Cultural fit entails more than just being courteous or getting along with coworkers. Identifying individuals who align with the company's values, beliefs, and work style is essential.

Employers frequently examine candidates cultural fit during behavioral interviews by probing their work habits, communication style, and problem-solving approach. These inquiries may include the following:

- What strategy do you use for teamwork and collaboration?

- What drives you to come to work every day?

- How do you handle disagreements with coworkers or managers?

- Describe your perfect workplace or environment where you would feel most productive.

- How do you strike a balance between your professional and personal lives?

- Tell me about when you had to adjust to a new working environment.

- To prepare for inquiries on cultural fit, learn about the company's goal, values, and work culture. Examine the company's website, social media accounts, and news stories. You may also inquire with current or previous workers about their job experiences at the organization.

- Being honest and sincere in your comments during the interview is critical. There may be better matches if you feel like you need to fit in with the company's culture. Nonetheless, if you believe you are culturally compatible with the organization, emphasize concrete examples from your previous experiences that indicate your ability to work successfully in a similar atmosphere.

Understanding the significance of cultural fit and preparing for relevant questions will help you succeed throughout the behavioral interview process. The STARR approach, a tried-and-true methodology for constructing replies to behavioral interview questions, will be covered in the next section.

Utilizing the STARR method when answering questions

When it comes to behavioral interview preparation, it is essential to have a method for structuring your responses. This approach will not only help you deliver comprehensive and insightful answers but will also demonstrate your clarity of thought, ability to self-reflect, and organizational skills. This section will discuss one such strategy: the STARR method.

The STARR method is a highly effective framework for organizing your responses during behavioral interviews. It helps in narrating your professional experiences by focusing on specific situations you've encountered, tasks you've handled, actions you've taken, and the resulting outcomes, all while reflecting on what you've learned from these experiences. The acronym **STARR** stands for the following:

- **Situation**: Explain the context of your circumstance. This should contain information about where you worked, whom you worked with, and other pertinent information.

- **Task**: Explain the assignment or difficulty that you were given. What was the goal or objective you were attempting to accomplish?

- **Action**: Explain the steps you followed to complete the work or solve the problem. What steps did you take, and why did you choose those actions?

- **Result**: Explain how your activities turned out. How did your actions affect the circumstance or task? Have you met your aim or objective?

- **Reflection**: Consider what you gained from your experience. What would you do differently? What worked well for you, and how can you build on that success in the future?

Using the STARR approach, you may frame your replies to behavioral interview questions in a clear and succinct manner. This may help you present relevant and exciting examples from your previous experiences and illustrate your talents and abilities in an easy-to-follow way for the interviewer.

The following subsections show example interview questions with example responses broken down in the STARR format.

Example interview question #1

Describe when you had to resolve a conflict with a teammate.

STARR method:

Situation: I was collaborating on a team project with a coworker who had divergent ideas about how we should approach the assignment.

Task: Our assignment was to develop a presentation to share our findings with the rest of the team.

Action: I arranged a meeting with my coworker to discuss our differences and find a solution to which we could agree. We listened to one another's viewpoints, identified areas of agreement, and formulated a plan incorporating our ideas.

Result: The rest of the team responded positively to our final presentation, and we were both pleased with the work we had accomplished together.

Reflection: I've learned that conflict resolution requires active listening, empathy, and a willingness to compromise.

Example interview question #2

Describe a time when you were required to manage multiple priorities.

STARR method:

Situation: I was working on a project with multiple components, each with its deadline and priority level.

Task: My responsibility was to ensure that all project components were completed on time and to a high standard.

Action: I created a detailed project plan outlining each component, its deadline, and the necessary resources for completion. I then prioritized each element based on its significance and urgency and delegated tasks to the appropriate team members.

Result: Despite the task's complexity, we completed the project on time and to a high standard.

Reflection: I learned that effective time management and prioritization are essential skills for managing complex projects and that delegation and collaboration can help ensure that all tasks are completed on time and to a high standard.

Example interview question #3

Describe when you were required to troubleshoot a technical issue.

STARR method:

Situation: I was tasked with extracting, transforming, and loading data from multiple sources into a centralized database for a data engineering project.

Task: My assignment was to investigate a technical problem causing the ETL process to fail.

Action: I examined the error logs to determine the cause of the issue and then collaborated with the database administrator and other team members to develop a solution. We identified an error in the ETL code as the cause of the failure, and by modifying the code, we resolved the issue.

Result: The ETL process was successful, and the data was successfully loaded into the centralized database.

Reflection: I learned the importance of collaboration and communication when resolving technical issues and the significance of paying close attention to error logs.

Example interview question #4

Describe a time when you were required to implement a data governance policy.

STARR method:

Situation: I worked for a financial services firm subject to stringent data privacy regulations.

Task: My responsibility was to implement a data governance policy to ensure compliance with these regulations.

Action: I collaborated with key stakeholders from across the organization to identify data elements that required protection and to develop policies and procedures to ensure these data elements' confidentiality, availability, and integrity. I also designed a training program to inform employees about the significance of data governance and the steps required to comply with the policy.

Result: The data governance policy was implemented successfully, and the organization could demonstrate compliance with applicable regulations.

Reflection: I have learned the significance of stakeholder engagement and education when implementing data governance policies and the need for continuous monitoring and evaluation to ensure continued compliance.

Example interview question #5

Describe a time when you had to optimize the efficiency of a data pipeline.

STARR method:

Situation: I was working on a data engineering project that required the daily processing of massive amounts of data.

Task: My responsibility was to optimize the data pipeline to decrease processing time and improve efficiency.

Action: I examined the existing pipeline architecture and identified improvement opportunities. Then, I collaborated with the development team to implement pipeline modifications, including code optimization and parallel processing techniques to increase throughput. I also collaborated with the infrastructure team to ensure that sufficient resources supported the pipeline's increased workload.

Result: The data pipeline was successfully optimized, resulting in a 50% reduction in processing time.

Reflection: I learned the significance of understanding the underlying architecture of data pipelines and the need for collaboration between data engineers and other development team members to optimize performance.

While adopting the STARR approach, it is critical to concentrate on the most relevant data and keep your response brief. Avoid using highly technical or jargon-heavy language and emphasize your unique position.

In the next section, we will explore some of the most frequently encountered questions in data engineering interviews. We'll provide sample responses to these questions, structured using the STARR method so that you can see them in action and gain valuable insights into how you might formulate your own impactful answers. So, let's delve deeper and get you ready for your upcoming interviews!

Reviewing the most asked interview questions

Data engineering interviews typically involve technical and behavioral questions to assess a candidate's technical knowledge, problem-solving skills, and ability to work effectively in a team.

The best approaches to answering these questions involve being specific about your experiences, highlighting your role in particular projects or tasks, discussing any challenges you faced and how you overcame them, and providing metrics that showcase your accomplishments.

Here are 10 of the most asked questions in a data engineering interview, along with how you might effectively respond to give better context:

- *Question 1*: What experience do you have with ETL processes and data pipelines?

  I have extensive experience with ETL processes and data pipelines. In my previous role, I was responsible for building a data pipeline that extracted data from various sources, transformed it, and loaded it into a centralized database. I used tools such as Apache NiFi and Apache Airflow to build and manage the pipeline. One of the main challenges I faced was dealing with data inconsistencies across different sources. I developed a series of data validation checks and implemented data cleansing techniques to overcome this.

- *Question 2*: Can you describe your experience with distributed systems and big data technologies?

  I have experience working with distributed systems and big data technologies such as Apache Hadoop, Apache Spark, and Apache Kafka. In a previous project, I was responsible for building a distributed data processing system that analyzed large volumes of streaming data. I used Apache Kafka as a message queue and Apache Spark for real-time processing to accomplish this. One of the main challenges I faced was optimizing the system's performance. I implemented several optimization techniques to overcome this, including data partitioning and parallel processing.

- *Question 3*: How do you ensure data quality and accuracy in your work?

  I highly emphasize ensuring data quality and accuracy in my work. I use techniques and tools such as data validation checks, data profiling, and data cleansing to accomplish this. For example, in a previous project, I developed a series of automated data validation checks that were run daily. These checks verified that the data in our database was consistent, accurate, and up to date.

- *Question 4*: Can you describe your experience with cloud-based data engineering platforms?

  I have experience working with several cloud-based data engineering platforms, including **Amazon Web Services** (**AWS**) and **Google Cloud Platform** (**GCP**). In a previous project, I used AWS to build a data pipeline that extracted data from multiple sources, transformed it, and loaded it into a centralized database. I used several AWS services, such as S3, Lambda, and Glue, to build and manage the pipeline. One of the main advantages of using a cloud-based platform was the ability to scale the pipeline to accommodate increased data volumes easily.

- *Question 5*: What experience do you have with real-time data processing and streaming technologies?

  I have experience working with several real-time data processing and streaming technologies, such as Apache Kafka and Apache Spark Streaming. In a previous project, I built a distributed data processing system that analyzed large volumes of streaming data in real time. I used Apache Kafka as a message queue and Apache Spark Streaming for real-time processing. One of the main challenges I faced was ensuring that the system could handle the high volume of incoming data. To overcome this, I implemented several optimization techniques, such as data partitioning and caching.

- *Question 6*: Can you describe a time when you had to work with a team to deliver a data engineering project?

  In a previous project, I worked with a team to build a data pipeline that extracted data from multiple sources, transformed it, and loaded it into a centralized database. I was responsible for the ETL process and collaborated closely with team members responsible for data modeling and database design. One of the main challenges we faced was ensuring that the pipeline was scalable and could accommodate future data growth. We used several optimization techniques, such as data partitioning and parallel processing, to overcome this.

- *Question 7*: How do you handle missing or corrupt data in a data pipeline?

  If missing or corrupt data is detected in a data pipeline, my approach is first to identify the root cause of the issue. Once the root cause has been identified, I develop a plan to address the issue. This may involve implementing data validation checks to prevent future occurrences of the problem, implementing data cleansing techniques to correct the issue, or working with upstream data providers to address the issue at the source. For example, in a previous project, we detected missing data in one of the data sources we used for the ETL process. After identifying the root cause, we worked with the upstream data provider to correct the issue at the source and implemented additional data validation checks to prevent future occurrences.

- *Question 8*: Can you describe your experience with data warehousing and **business intelligence (BI)** tools?

  I have experience working with data warehousing and BI tools such as Snowflake and Tableau. In a previous project, I built a data warehouse that consolidated data from multiple sources and provided a centralized source of truth for reporting and analysis. I used Snowflake to develop and manage the data warehouse and Tableau for reporting and visualization. One of the main challenges I faced was ensuring that the data warehouse could accommodate future data growth. I used several optimization techniques, such as data partitioning and compression, to overcome this.

- *Question 9*: How do you ensure data privacy and security in your work?

  I highly emphasize ensuring data privacy and security in my work. I use several techniques and tools to accomplish this, such as data encryption, access control, and audit logging. For example, in a previous project, I implemented data encryption to protect sensitive data during transit and at rest. I also implemented access controls to ensure only authorized personnel could access sensitive data and audit logging to track data access and changes.

- *Question 10*: Can you walk me through your approach to troubleshooting a data engineering issue?

  When troubleshooting a data engineering issue, my approach is first to identify the symptoms of the issue and gather relevant information, such as log files and error messages. Once I have a clear understanding of the issue, I develop a hypothesis about the root cause of the issue and test the theory by performing additional analysis or testing. If the idea is confirmed, I will create a plan to address the issue, which may involve implementing a workaround or a more permanent solution. For example, in a previous project, we encountered an issue with a data pipeline that was causing a high volume of errors. After gathering relevant information and developing a hypothesis, we confirmed that a bug in the ETL code caused the issue. We implemented a temporary workaround to address the issue and worked on a permanent solution to fix the bug.

Summary

Preparing for behavioral interviews is critical to any job search, especially for data engineering positions. Candidates can better prepare for success by understanding the types of behavioral questions commonly asked in data engineering interviews and the skills that are measured during these interviews.

This chapter discussed six main types of behavioral questions, including those that assess cultural fit and the STARR method for effectively answering behavioral questions. We also reviewed the top 10 most frequently asked questions in a data engineering interview and provided detailed explanations of how to answer them.

By employing these strategies and best practices, candidates can increase their chances of success in a data engineering interview and effectively demonstrate their skills and qualifications. As the demand for data engineering roles grows, being well prepared for these interviews is more crucial than ever. We hope this chapter has provided candidates with valuable insights and guidance.

In the next chapter, we will delve into specific Python skills that every aspiring data engineer should have under their belt.

Part 2:
Essentials for
Data Engineers Part I

In this part, we will begin to provide an overview of essential areas of Python, SQL, and databases.

This part has the following chapters:

- *Chapter 5, Essential Python for Data Engineers*
- *Chapter 6, Unit Testing*
- *Chapter 7, Database Fundamentals*
- *Chapter 8, Essential SQL for Data Engineers*

5

Essential Python for Data Engineers

Finding your way through the data engineering interview process can be challenging, particularly when it comes to showcasing your technical expertise. Python is frequently the preferred language for data engineering tasks because of its ease of use, readability, and rich library support. A solid understanding of Python is essential for anyone working with data **Extraction, Transformation, and Loading** (**ETL**) procedures or developing intricate data pipelines.

This chapter aims to give you the Python knowledge you need to succeed in a data engineering position. We'll begin by discussing the fundamental Python skills that each data engineer should be familiar with. We'll then get into more complicated subjects that will make you stand out from other candidates. We'll finish the chapter with some technical interview questions that assess your knowledge of Python in the context of data engineering.

In this chapter, we will cover the following topics:

- Must-know foundational Python skills
- Must-know advanced Python skills
- Technical interview questions

Must-know foundational Python skills

In this section, we concentrate on the fundamental Python concepts necessary for data engineering. This entails being familiar with the syntax of Python as well as basic data structures such as lists, tuples, and dictionaries. We'll look at how to use control structures such as conditional statements, loops, and functions, as well as how to create and use them. The importance of Python's basic built-in functions and modules will be emphasized, along with its role in creating effective, modular programming.

We'll finish up by discussing file **input/output** (**I/O**) operations, which are crucial for processing data. The overview of these crucial Python foundations in this section will help you get ready for and ace your data engineering interview. It's not a Python course, but rather a review of the fundamental abilities a data engineer needs.

In the upcoming subsections, the foundational skills have been broken down into five sections.

SKILL 1 – understand Python's basic syntax and data structures

Python is a fantastic choice for novices because its syntax is straightforward and quick to grasp. The syntax of the language must first be understood, including its basic concepts of variables, operators, and control structures such as conditional statements and loops. Every data engineer should be familiar with basic Python concepts such as lists, tuples, and dictionaries.

Statements in Python code are executed by the interpreter line by line. These statements may contain variables, operators, control structures, and functions. Operators manipulate data values mathematically or logically, whereas variables store data values. Control structures such as `if-else` statements and loops are used to regulate the program's flow. Similar pieces of code are grouped together into reusable chunks using functions.

Success in data engineering requires knowledge of Python's data structures in addition to its syntax. Lists, tuples, and dictionaries are examples of data structures that store a collection of data values. Lists and tuples are both ordered collections of values; however, tuples are immutable and lists are not. A unique key can be used to store and retrieve data from collections of key-value pairs, known as dictionaries.

An important component of Python syntax is indentation. Python, unlike many other programming languages, uses indentation rather than curly brackets or other conventions to separate code chunks. This means that for Python code to run properly, sufficient indentation is necessary. Learning Python's syntax and data structures needs lots of practice in building and running programs. Python tutorials and exercises are offered by a number of online groups and resources such as the Data in Motion platform and the **Data Career Academy** (**DCA**). Building a strong Python foundation can also be facilitated by reading through the documentation and playing with various syntaxes and data structures.

SKILL 2 – understand how to use conditional statements, loops, and functions

Conditional statements, such as `if-else` statements, enable you to run different code blocks based on specific conditions. For example, you could use an `if-else` statement to determine whether a variable is greater than a specific value and then execute different code blocks based on the result.

`for` and `while` loops are used to iterate over a set of values or to perform a repetitive task. A `for` loop, for example, could be used to iterate over the elements of a list and perform a calculation on each element.

Functions are used to organize similar code into reusable blocks. This facilitates the organization and modularization of your code. For example, you could create a function to perform a specific data transformation task and call it from multiple places in your code.

Understanding best practices for writing efficient and readable code and how to use these constructs is critical. For example, using descriptive variable names and commenting on your code can make it easier to understand and maintain.

To become acquainted with conditional statements, loops, and functions in Python, it is necessary to write code and experiment with various syntaxes and techniques. Online resources such as Python documentation, interactive tutorials, and coding challenges can aid in developing these skills. Reading through Python code written by experienced programmers can also help you learn best practices and develop your programming style.

SKILL 3 – be familiar with standard built-in functions and modules in Python

Python provides several built-in functions and modules that can simplify programming tasks and help optimize code. As a data engineer, it is essential to be familiar with these built-in functions and modules, as they can be used to perform everyday data manipulation tasks and improve the performance of your code.

Some standard built-in functions in Python include the following:

- `print()`: Used to display output to the console
- `len()`: Used to determine the length of a string or list
- `range()`: Used to generate a sequence of numbers

Python also includes several built-in modules that can be used for more advanced programming tasks, such as data analysis and visualization. Some commonly used modules include the following:

- `math`: Provides functions for mathematical operations, such as trigonometry and logarithms
- `random`: Provides functions for generating random numbers
- `datetime`: Provides functions for working with dates and times
- `os`: Provides functions for interacting with the operating system, such as creating and deleting files

In addition to these built-in functions and modules, Python has a large ecosystem of third-party libraries and modules that can be used for specific tasks, such as data analysis or **machine learning** (**ML**). These libraries, such as NumPy and pandas, can significantly simplify complex tasks and improve the efficiency of your code.

To become familiar with these built-in functions, modules, and third-party libraries, it is essential to practice working with them and experimenting with different use cases. Reading through code written by experienced Python programmers and analyzing how they use built-in functions and modules can provide valuable insights and best practices for improving your code.

SKILL 4 – understand how to work with file I/O in Python

Data engineers often work with large datasets that are stored in files. Python provides several built-in functions for file I/O operations essential for working with these datasets.

Two primary modes for file I/O in Python are reading and writing. When reading from a file, the contents of the file are loaded into memory and made available for manipulation. When writing to a file, data is written to the file for storage or later retrieval.

To open a file in Python, you use the `open()` function, which takes two arguments: the file's name and the mode in which you want to open it (read or write). For example, to open a file for reading, you would use the following code:

```
file = open('filename.txt', 'r')
```

Once you have opened a file, you can read or write data using a variety of built-in functions. For example, to read the contents of a file, you can use the `read()` function, which reads the entire contents of the file as a string. To read a single line from a file, you can use the `readline()` function.

To write data to a file, you can use the `write()` function, which writes a string to the file. You can use the `writelines()` function to write multiple lines to a file.

It is important to note that when working with files in Python, you must permanently close the file when you are finished working with it. This is typically done using the `close()` function.

To become comfortable working with file I/O in Python, it is important to practice reading and writing files using different techniques and syntax. Online resources such as Python documentation and tutorials can help build these skills. Additionally, working on small projects that involve reading and writing files can help you gain practical experience and improve your understanding of how file I/O works in Python.

Now that we've covered the fundamentals of Python, we'll go into more complicated topics to close the knowledge gap between fundamental knowledge and the level of expertise anticipated by a skilled data engineer.

SKILL 5 – functional programming

A programming paradigm known as **functional programming** avoids changing state or mutable data and treats computation as the evaluation of mathematical functions. Functional programming can be beneficial for data engineering because it makes testing, debugging, and parallelization simpler.

Despite not being a purely functional language, Python offers many features that support functional programming, and being familiar with these features can be very helpful in a data engineering role.

The use of higher-order functions such as `map()`, `filter()`, and `reduce()` is one of the most fundamental ideas in functional programming in Python. Without having to create explicit loops, you can use these functions to perform operations on lists and other iterable data structures. For instance, using `map()` to apply a function to every element in a list simplifies and improves the readability of data transformation tasks.

Another feature of Python's functional programming language is lambda functions, also known as anonymous functions. The `lambda` keyword is used to define these short, inline functions, which are frequently used for quick, straightforward operations inside of higher-order functions. To square each item in a list, for instance, you could use a lambda function inside a `map()` call.

Pythonic methods for data transformation or filtering include list comprehensions and generator expressions. You can accomplish the same goals as loops or higher-order functions using their more declarative approach to data manipulation, but with more readable syntax. Additionally, pure functions, which have sole dependence on their input and a lack of any side effects, are encouraged by functional programming. Pure functions are a good choice for data engineering tasks that demand dependability and repeatability because they are simpler to test and debug.

Last but not least, Python provides modules such as `functools` that offer sophisticated functional programming tools such as decorators for function composition or memorization. These can be especially helpful for enhancing performance in tasks that require lots of data. You can create Python code that is more effective, spick-and-span, and easy to maintain by using functional programming techniques. Not only will knowing these ideas improve your Python programming, but it will also make you a better data engineer.

Must-know advanced Python skills

In addition to the fundamental Python skills covered in the preceding section, data engineers should be familiar with several advanced Python concepts and techniques. Examples are **object-oriented programming** (**OOP**), advanced data structures, and well-known libraries and frameworks for data analysis and visualization.

In the following subsections, we will review the advanced Python skills required for data engineering interviews.

SKILL 1 – understand the concepts of OOP and how to apply them in Python

The foundation of the OOP programming paradigm is the concept of objects, which can hold data and allow for the coding of data manipulation. When developing complex software systems for data engineering applications, OOP is a powerful technique. Python uses classes to create objects that csn be created and used to do certain tasks. A class is a blueprint that details the characteristics and capabilities of an item. When you instantiate an object, a new instance of the class is produced with all of its features.

Inheritance is a crucial concept in OOP because it allows developers to create new classes that inherit the attributes and functions of preexisting ones. Use the syntax given next to define a subclass in Python that derives from a superclass:

```
class SubClass(SuperClass):
    # Class definition goes here
```

Polymorphism, which allows you to use objects of different types interchangeably, is another important concept in OOP. Python achieves polymorphism through duck typing, meaning an object's type is determined by its behavior rather than its class.

Working with classes and objects and experimenting with different inheritance and polymorphism scenarios is essential for becoming comfortable with OOP concepts and applying them in Python. Online resources, such as Python documentation and tutorials, can aid in developing these skills. Working on small OOP-related projects can also help you gain practical experience and improve your understanding of how OOP works in Python.

SKILL 2 – know how to work with advanced data structures in Python, such as dictionaries and sets

While lists and tuples are the most commonly used data structures in Python, several more advanced data structures can be helpful for certain types of data manipulation tasks. Dictionaries and sets are examples of these.

A dictionary is an unsorted set of key-value pairs. Each key-value pair is saved in the dictionary as an item, and the key is used to access the corresponding value. Dictionaries are useful for storing data, making quick lookups based on a unique key possible. Dictionaries are defined in Python using curly braces ({ }) and colons (:), as shown in the following example:

```
my_dict = {'key1': 'value1', 'key2': 'value2', 'key3': 'value3'}
```

A set is an unordered collection of unique elements. Sets help perform set operations, such as union and intersection. In Python, sets are defined using curly braces ({ }) and commas (,), as shown in the following example:

```
my_set = {1, 2, 3, 4, 5}
```

Understanding the various methods and operations available for manipulating dictionaries and sets to work effectively with these data structures is critical. For example, the update() and pop() methods can add and remove items from a dictionary. You can use the union() and intersection() methods to perform operations on sets such as unions and intersections.

To become acquainted with dictionaries and sets in Python, practicing using these data structures in various scenarios and experimenting with different methods and operations is necessary. Online resources, such as Python documentation and tutorials, can aid in developing these skills. Working on small projects involving dictionaries and sets can also help you gain practical experience and better understand how these data structures work in Python.

SKILL 3 – be familiar with Python's built-in data manipulation and analysis libraries, such as NumPy and pandas

NumPy and pandas are two widespread Python data manipulation and analysis libraries. NumPy handles large, multidimensional arrays and matrices, whereas pandas handles data manipulation and analysis, including tools for reading and writing data to and from various file formats.

NumPy is a must-have Python library for scientific computing. It includes a multidimensional array object and tools for manipulating these arrays. The `ndarray` (n-dimensional array) object, which provides efficient storage and manipulation of large arrays of data, is at the heart of NumPy's functionality. NumPy also includes trigonometric and logarithmic functions for working with arrays.

pandas is a data manipulation and analysis library that provides high-performance data manipulation and analysis tools. It includes data structures such as the `Series` and `DataFrame` objects for efficiently storing and manipulating large datasets. pandas also has tools for reading and writing data to and from various file formats, including CSV and Excel.

To become acquainted with NumPy and pandas, working with these libraries and experimenting with various use cases is necessary. Online resources, such as NumPy and pandas documentation and tutorials, can aid in developing these skills. Working on small projects involving data manipulation and analysis can also help you gain practical experience and improve your understanding of how these Python libraries work.

SKILL 4 – understand how to work with regular expressions in Python

Regular expressions (**regexes**) are a powerful text-processing tool. The `re` module in Python supports regexes by providing functions for searching, matching, and manipulating text data based on specific patterns.

Regexes describe patterns in text data, such as the presence of specific words or character patterns. A regex, for example, can be used to search for all instances of a particular word in a large text file. Regexes can also perform more complex pattern-matching tasks, such as finding email addresses or phone numbers in a block of text.

Regexes in Python are defined using a unique syntax that allows you to describe the pattern you're looking for. To find all instances of the word `"data"` in a string, for example, use the following regex:

```
import re

text = "This is some data that I want to search for"
pattern = "data"

matches = re.findall(pattern, text)
```

In this example, the `findall()` function searches the text variable for all instances of the `pattern` regex and returns a list of all matches.

It is critical to practice using different regexes and experiment with different search and match functions to become comfortable working with regexes in Python. Working on small projects that use regexes can also help you gain practical experience and better understand how regexes work in Python.

SKILL 5 – recursion

Programming recursion is the process of having a function call itself in order to solve a problem. Recursion has practical significance for data engineers even though it may initially seem like a topic more related to computer science theory or algorithmic difficulties. Recursion can be especially helpful for tasks involving nested or hierarchical data structures, such as JSON objects, XML files, or database schemas that resemble trees. Additionally, it may be crucial for algorithms that call for backtracking, divide-and-conquer tactics, or depth-first search.

You might run into situations in data engineering where you need to navigate a nested data structure in order to extract, transform, or load data. For instance, you might need to flatten a JSON object into a tabular form for analysis if it contains nested arrays or dictionaries. Such nested structures can be gracefully handled by recursive functions, which divide the problem into more manageable, smaller subproblems.

Graph algorithms are one more common application of recursion in data engineering. Recursive algorithms such as depth-first search can be extremely helpful for tasks such as locating the shortest path or spotting cycles, regardless of whether you're working with social network graphs, dependency graphs in software, or any other type of interconnected data.

Recursion can, however, be resource-intensive if not implemented carefully, particularly in terms of memory usage. You can change Python's default recursion depth limit, but doing so has drawbacks. Understanding the trade-offs and knowing when to use iterative solutions in place of others can, therefore, be very important.

Recursion is also well suited to parallelization, which is frequently necessary in tasks requiring scale in data engineering. Recursive functions can occasionally be divided into separate subtasks that can be carried out in parallel, enhancing performance.

In conclusion, recursion is a potent tool that can make difficult data engineering problems simpler, improving the readability and maintainability of your code. In addition to assisting you in technical interviews, mastering recursion will give you a flexible skill set for addressing a variety of data engineering challenges.

Now that you are familiar with essential Python concepts, we will next go over sample interview questions and answers.

Technical interview questions

Technical interviews for data engineering positions often include questions related to Python programming concepts and techniques and broader technical concepts related to data engineering. In the following sections, we will see 15 difficult technical interview questions related to Python, data engineering, and general technical concepts.

Python interview questions

The following is a list of Python interview questions:

- *Question 1*: What is a descriptor in Python?

  Answer: A descriptor is a particular Python object that allows you to define how an attribute is accessed or modified in a class. Descriptors are commonly used to define properties, enabling you to control how attributes are accessed and modified.

- *Question 2*: How do you handle circular imports in Python?

  Answer: Circular imports occur when two or more modules import each other. To handle circular imports in Python, you can use several techniques, such as importing the module at the function level, using a separate module for the shared code, or using a lazy import.

- *Question 3*: What are the differences between a shallow copy and a deep copy in Python?

  Answer: A shallow copy creates a new object that points to the original object's memory location. In contrast, a deep copy constructs a new object with its memory that contains a copy of the original object's data. A shallow copy is valid when creating a new object that refers to the same data as the original object. In contrast, a deep copy is applicable when creating a completely separate copy of the original data.

- *Question 4*: What is the difference between a list comprehension and a generator expression in Python?

  Answer: A list comprehension is a shorthand syntax for creating a list from an iterable, while a generator expression is a shorthand syntax for creating a generator from an iterable. List comprehensions create a list in memory, while generator expressions create a generator object that generates values on the fly.

- *Question 5*: What is the difference between bound and unbound methods in Python?

  Answer: A bound method is a method that is bound to an instance of a class, while an unbound method is a method that is not bound to an instance of a class. Bound methods can access the instance's attributes and methods, while unbound methods cannot.

Data engineering interview questions

In this section, we'll go over typical data engineering interview questions. This section aims to give you a realistic idea of what to anticipate in interviews and how to communicate your knowledge correctly. We hope to inspire confidence and improve your readiness for the dynamic field of data engineering by breaking down complex topics into a question-and-answer style. Let's look at some real-world data scenarios and unravel their complexities with some sample questions and answers, as follows:

- *Question 1*: What is lambda architecture, and how does it work?

  Answer: A lambda architecture is a data processing architecture that combines batch processing and stream processing to provide both real-time and historical views of data. A lambda architecture processes incoming data in parallel using batch processing and stream processing systems and merges the results to give a complete data view.

- *Question 2*: What are the differences between a data lake and a data warehouse?

  Answer: A data lake is a large, centralized repository for storing raw data from multiple sources. In contrast, a data warehouse is a repository for storing structured and processed data that has been transformed and cleaned for analysis. Data lakes typically store large volumes of unstructured or semi-structured data, while data warehouses store structured data optimized for analytics.

- *Question 3*: How do you optimize a database for write-heavy workloads?

  Answer: Optimizing a database for write-heavy workloads typically involves sharding, partitioning, and clustering to distribute the workload across multiple nodes and using indexing, caching, and batching to improve the efficiency of write operations.

- *Question 4*: What is the difference between a batch processing system and a stream processing system?

  Answer: A batch processing system processes data in discrete batches, while a stream processing system processes data in real time as it arrives. Batch processing systems are typically used for analyzing historical data, while stream processing systems are used for processing real-time data.

- *Question 5*: What are the challenges of working with distributed systems, and how can they be addressed?

  Answer: Challenges of working with distributed systems include network latency, data consistency, and fault tolerance. These challenges can be addressed through data replication, partitioning, and load-balancing techniques.

General technical concept questions

We now shift our focus from Python and data engineering specifics to broader technical principles that can be asked about in interviews. A thorough review of these common technical questions is provided in this section. Here's a list of typical interview questions:

- *Question 1*: What is the difference between a mutex and a semaphore?

  Answer: A mutex locking mechanism ensures that only one thread can access a shared resource at a time. At the same time, a semaphore is a signaling mechanism that allows multiple threads to access a shared resource concurrently up to a specified limit. Mutexes are typically used to protect critical sections of code, while semaphores control access to shared resources.

- *Question 2*: What is the CAP theorem, and how does it relate to distributed systems?

  Answer: According to the CAP theorem, a distributed system cannot concurrently offer the three guarantees of consistency, availability, and partition tolerance. Consistency in a distributed system is the idea that all nodes concurrently see the same data, while availability is the idea that the system keeps working even if some nodes fail. The concept of partition tolerance states that the system can still function in the event of a network partition. The CAP theorem is often used to guide distributed systems' design and implementation.

- *Question 3*: What is the difference between monolithic and microservices architecture?

  Answer: A monolithic architecture is a software architecture in which all of the components of an application are tightly coupled and deployed as a single unit. In contrast, a microservices architecture is a software architecture in which an application is broken down into a collection of small, loosely coupled services that can be developed and deployed independently. Microservices architectures offer greater flexibility and scalability than monolithic architectures but require more complex infrastructure and deployment processes.

- *Question 4*: What is the difference between **Atomicity, Consistency, Isolation, Durability (ACID)** and **Basically Available, Soft state, Eventually consistent (BASE)** consistency models?

  Answer: ACID is a database consistency model that provides strong guarantees around data consistency and integrity. BASE is a consistency model that sacrifices strong consistency in favor of availability and partition tolerance. BASE is often used in distributed systems where strong consistency is difficult to achieve.

- *Question 5*: What are the different types of joins in SQL, and how do they work?

  Answer: The various join types in SQL include full outer join, left outer join, right outer join, and inner join. All rows from the left table and all matching rows from the right table are returned by a left outer join, all rows from the right table and all matching rows from the left table are returned by a right outer join, and all rows from both tables are returned by a complete outer join with `null` values in the columns where no match was found. On the basis of common columns, joins are used to aggregate data from many tables.

Summary

We covered a lot of the Python knowledge required for aspiring data engineers in this chapter. We started by building a solid foundation with fundamental Python knowledge by going over syntax, data structures, and control flow components such as conditional statements, loops, and functions. We also gave you a brief introduction to Python's standard built-in modules and functions, which are useful for a variety of data engineering tasks. We also looked at the idea of functional programming in Python, emphasizing its advantages for producing effective, orderly, and maintainable code.

We talked about OOP principles and how they can be used in Python to create modular and reusable code as we progressed into advanced Python skills. In order to effectively handle complex data, advanced data structures such as dictionaries and sets were also covered. Given their widespread usage in data engineering tasks, familiarity with Python's built-in data manipulation and analysis libraries, such as NumPy and pandas, was stressed. We also discussed how crucial it is to comprehend regexes for text editing and data cleansing. We concluded by discussing the idea of recursion and outlining its application in data engineering, particularly when working with nested or hierarchical data structures.

We now go on to the next chapter, *Unit Testing*. The importance of testing will be discussed in the next chapter, emphasizing Python unit testing techniques to ensure your data pipelines and applications are reliable and error-free in addition to being functional. Let's continue and expand on the solid Python foundation we've established.

6

Unit Testing

Unit testing is a critical part of a data engineering workflow. It allows the data engineer to test individual units in their code. A unit is any independent piece of code and can be functions, modules, or classes. They will test each unit in isolation to ensure the intended output is returned and functions properly.

Unit testing allows the data engineer to find and fix potential bugs earlier in the development process and also protects their code from breaking when changes are done or their code becomes more complex. It is also more efficient than manual testing.

In this chapter, we will cover fundamental-to-intermediate unit testing concepts in Python you should know for an interview.

In this chapter, we will cover the following topics:

- Fundamentals of unit testing
- Must-know intermediate unit testing skills
- Unit testing interview questions

Fundamentals of unit testing

In data engineering, unit tests are essential to validate data transformations, computations, and other data manipulation tasks, ensuring that every function or method in your pipeline works correctly. These tests enable you to catch and correct errors early in development, saving time and resources in the long run.

In the upcoming sections, we will explore the fundamentals of unit testing, discussing its principles, advantages, and how to implement it effectively in your code. We will cover various aspects such as writing testable code, structuring your tests, and using testing frameworks in Python, such as `pytest` and `unittest`. Whether you are new to the concept or looking to enhance your testing skills, this section will serve as a comprehensive guide to mastering unit testing in data engineering.

Importance of unit testing

Unit testing is a fundamental component of creating reliable, maintainable, and scalable data pipelines. It is also integrated within continuous integration and development. As a data engineer, you want to be able to create code that can withstand future changes and potential complexities in production. Unit tests ensure your functions will work as expected as complex changes or inputs in production occur.

Testing units in isolation also ensures the ability to detect errors and bugs early in the development process. Unit testing allows you to implement tests that act as a guardrail for different potential and predicted edge cases.

Unit testing frameworks in Python

In this section, we'll look at many of the most popular options for putting Python's code through its paces. These include but are not limited to `pytest`, an advanced and feature-rich library with widespread usage in the industry, and `unittest`, Python's in-built testing library.

To maximize your testing efficiency, you should familiarize yourself with the specifics of each of these frameworks. For instance, to facilitate test automation, `unittest` provides a stable groundwork for test structuring via test cases, test suites, and a test runner. On the other hand, `pytest` provides advanced testing capabilities including reusable fixtures, parameterized tests, and comprehensive error reports.

We will now look at code examples where we will incorporate both the `unittest` and `pytest` frameworks.

`unittest` is a built-in testing framework within Python's standard library that provides tools for performing, organizing, and running unit tests.

Here is an example of using `unittest` for a function under the filename `math_operation.py`:

1. Create a function called `add` that will be tested under the filename `math_operation.py`:

    ```
    # math_operations.py

    def add(a, b):
        return a + b
    ```

2. Create a separate file called `test_math_operations_unit.py` that will perform unit tests on the `add` function. Each test is defined by different functions that evaluate results from different scenarios:

    ```
    # test_math_operations.py

    import unittest
    from math_operations import add
    ```

```
class TestMathOperations(unittest.TestCase):

    def test_add_positive_numbers(self):
        self.assertEqual(add(2, 3), 5)

    def test_add_negative_numbers(self):
        self.assertEqual(add(-2, -3), -5)

    def test_add_mixed_numbers(self):
        self.assertEqual(add(-2, 3), 1)
        self.assertEqual(add(2, -3), -1)

if __name__ == '__main__':
    unittest.main()
```

pytest is a third-party framework that has more robust features such as fixtures and parameterization. Fixtures allow you to feed data or input to unit tests. Parameterization allows you to run the same unit test at once with different sets of inputs.

Here is an example of using pytest to perform a unit test on the previous add function:

1. Install pytest, as follows:

    ```
    pip install pytest
    ```

2. Create a separate file called test_math_operations_pytest.py, which will perform unit tests on the add function. Each test is defined by different functions that evaluate results from different scenarios:

    ```
    # test_math_operations.py

    from math_operations import add

    def test_add_positive_numbers():
        assert add(2, 3) == 5

    def test_add_negative_numbers():
        assert add(-2, -3) == -5

    def test_add_mixed_numbers():
        assert add(-2, 3) == 1
        assert add(2, -3) == -1
    ```

While both frameworks have their merits, the choice between unittest and pytest often comes down to your project's requirements and personal preference. When used effectively, both are potent instruments that can improve the robustness and dependability of your data pipelines.

Having examined these examples, we are better suited to write, structure, and execute unit tests in Python code. As we progress, we will apply these skills to more complex areas of data engineering, continuing to build upon the solid foundation that unit testing provides.

Process of unit testing

Unit testing, despite its importance, is not a procedure that can be performed without a systematic approach. It involves multiple steps, including gaining a comprehension of the unit's intended functionality, writing and executing tests, and integrating them into your development pipeline. Every data engineer must comprehend this procedure to ensure the development of efficient, robust, and trustworthy data pipelines.

In this section, we will examine the unit testing procedure in detail. We will provide a systematic approach to help you comprehend how to effectively plan, write, execute, and maintain unit tests.

When it comes to structuring and running unit tests, there is typically a three-step process is followed, also known as the three As, elaborated upon here:

- **Arrange**: Here, you set up your environment for the test and load any necessary data, modules, or other preconditions. Specifically, this can involve creating instances of the classes you need to test, defining input data, or setting up mock objects.

- **Act**: Here, you call the method or function to test functionality. This is a core part of the process when the unit is actually tested.

- **Assert**: Here, you verify the result of the previous *Act* step. This is where your test will let you know if it passes or fails. If it does fail, it will detail what the failure is and where it occurs in your code.

Here are the basic steps of unit testing broken down using the `unittest` framework:

1. Import `unittest`, like so:

   ```
   import unittest
   ```

2. Create your class of test cases, as follows:

   ```
   class MyTestCase(unittest.TestCase):
       def test_something(self):
           # test code
   ```

3. Fill your class with different test methods. The methods represent specific scenarios that you will test your function under:

   ```
   def test_something(self):
       # test code
   ```

4. Add a `unittest.main()` method so that all tests will run when the unit test is called to run:

```
if __name__ == '__main__':
    unittest.main()
```

After these steps are completed, you can run your unit tests in the command line by calling the filename where your tests are created. This is a basic example of how to run your unit tests, and it can be altered by implementing more advanced components such as fixtures or parameterized tests.

Now that you understand the process of running a unit test, we will introduce more intermediate concepts in the next section.

Must-know intermediate unit testing skills

While unit testing fundamentals provide an essential foundation, data engineering frequently necessitates a more profound comprehension and a broader set of testing skills. Complex, high-volume data transformations and operations need more sophisticated testing strategies to ensure your pipelines' robustness, dependability, and effectiveness.

We'll begin by discussing parameterized testing, a technique enabling you to execute the same test case with various input values, expanding your test coverage without duplicating code. Then, we will conduct performance and duress testing to ensure that your code can withstand the demands of actual data operations.

Parameterized tests

Parameterized tests allow you to run the same unit test but with multiple inputs of different data. This allows you to test different scenarios in less code rather than writing out multiple tests.

The following is an example of a parameterized test using the same `add` function we previously used.

Create a separate file called `test_math_operations_param.py`, which will perform unit tests on the `add` function using parameterized testing in `pytest`. We will input a list containing different tuples of integers:

```
# test_math_operations.py

import pytest
from math_operations import add

@pytest.mark.parametrize(
    "a, b, expected_sum",
    [
        (2, 3, 5),
        (-2, -3, -5),
```

```
        (-2, 3, 1),
        (2, -3, -1),
        (0, 0, 0)
    ]
)
def test_add(a, b, expected_sum):
    result = add(a, b)
    assert result == expected_sum
```

There is only one testing function, `test_add`, with an `assert` statement to evaluate the function using each tuple.

Performance and stress testing

Performance and stress testing are both non-functional testing methods. Performance testing measures how well your unit performs in normal conditions. It can also identify bottlenecks and areas in your code that are causing performance issues. This testing will ensure your code can handle expected load conditions before deployment.

Stress testing measures how well your code will handle unexpected loads and edge-case scenarios. Here, you can identify potential breaking points in your code and spot potential failures before deployment.

Various scenario testing techniques

While stress testing measures how your code will behave in extreme or unexpected circumstances, it is also important to test your code under various expected conditions. Examples include the following:

- **Happy-path testing**: Ensures your code functions as expected during all normal conditions with a valid input.

- **Boundary-value testing**: Ensures your code performs as expected with edge cases of inputs such as negative numbers, very large numbers, empty strings, or other extreme values.

- **Error-handling testing**: Ensures your code can handle invalid inputs and others that would result in predictable errors.

We are now well versed in primary intermediate unit testing skills in data engineering, including parameterized testing, performance and stress testing, and scenario testing. Complex as they may be these techniques are essential for ensuring the robustness, efficacy, and dependability of your data pipelines under a variety of conditions.

However, comprehension of these concepts is only half the battle. You must be able to communicate this understanding effectively in your interview. Moving on to the next section, we will apply the knowledge acquired thus far. The next section will provide examples of how concepts and techniques related to unit testing may be evaluated during an interview. We will cover various topics, from the fundamentals of unit testing to the intermediate skills we've just discussed.

Unit testing interview questions

Now that we have provided an overview of unit testing in Python, this section will provide example questions and answers to help prepare you for your interviews, as follows:

- *Question 1*: Explain the difference between unit testing and integration testing.

  Answer: Unit testing is testing individual components or functions in isolation. Integration testing focuses on testing how a section of code interacts within the whole system. Unit testing tests for proper functions for an individual component, while integration testing validates how well different components perform together.

- *Question 2*: How are assertions used in unit testing?

  Answer: Assertions are used to validate the expected behavior of a function. They compare the expected output and the actual output to evaluate for accuracy.

- *Question 3*: What are ways you can improve the performance of your unit tests?

  Answer: Minimize dependencies, utilize the `setUp()` and `teardown()` methods, optimize your code, and run tests in parallel.

- *Question 4*: What is continuous testing and why is it important in unit testing?

  Answer: Continuous testing runs your unit tests automatically. This will help reduce regression and identify issues early.

- *Question 5*: What are examples of external dependencies, and what are strategies to handle them in unit testing?

  Answer: Examples of external dependencies are APIs and databases. A strategy to minimize dependencies in your unit test is mocking. This allows you to use a simulated and predefined version of the data for your unit tests.

- *Question 6*: What is regression testing?

  Answer: Regression testing is the process of testing your code to ensure it still works as expected over time as code changes, bug fixes, and feature implementations occur.

- *Question 7*: What is the purpose of test fixtures?

  Answer: Test fixtures ensure your unit tests have consistent and stable environments and setup operations. This can mean consistent data, variables, or database connections. This allows for reliable and repeatable test results.

Summary

In this chapter, we explored the skill of unit testing, from its fundamental importance to mastering its practical application with Python's `unittest` and `pytest` frameworks. We expanded our skills by exploring intermediate concepts, such as parameterized, performance, and scenario testing, and concluded with a practical guide to tackling unit testing questions in job interviews.

As we transition into the next chapter, we'll move from the coding realm to the data world, focusing on essential knowledge of databases and their operation. This shift will further enrich your data engineering skills, providing the necessary tools to interact with, manipulate, and manage data effectively in your future roles. Let's continue our journey into the heart of data engineering.

7

Database Fundamentals

The database is your workstation, and you are the architect and keeper of the data as a data engineer. So, it's essential to thoroughly understand databases, how they work, and all of their subtleties. Beginning with the fundamental foundational concepts, we will gradually move on to more complicated features of databases in this chapter before dissecting these principles via the context of a data engineering interview.

This chapter will walk you through various topics, whether you're a novice just entering this industry or a seasoned professional trying to brush up on your knowledge. These include database types, normalization, and indexing fundamentals. In a series of common interview questions and responses, we will put the lessons gained to use. Your technical communication abilities will be improved in this segment, which is a crucial component of any data engineering interview. The chapter will conclude with a brief summary to help you remember the most important ideas.

In this chapter, we will review how data engineers use databases, which you should be familiar with prior to an interview, including the following:

- Must-know foundational database concepts
- Must-know advanced database concepts
- Technical interview questions

Must-know foundational database concepts

In this section, we will create the groundwork for your understanding of databases, which is a crucial component for any data engineer. We start with the fundamentals, such as comprehending the many types of databases, the normalization tenets, and the idea of indexing. These foundational elements will serve as the starting point for your exploration of the more complex world of data engineering. This part is the starting point for a more in-depth investigation of database systems, whether you are a beginner or an experienced professional reviewing the fundamentals. Let's begin by strengthening your fundamental database knowledge.

Relational databases

A relational database utilizes a relational model to store structured data into tables. Tables are collections of rows and columns. Each row in the table represents a single record of information, while each column represents attributes or fields. The following screenshot is an example of a simple SQL table with its rows and columns labeled:

Id	Name	Surname	Age
1	Jodie	Tucker	34
2	Jayden	Archer	56
3	Grace	Wheeler	18
4	Freddie	Humphries	56

Columns ←   Rows →

Figure 7.1 – Example of a SQL table highlighting the rows and columns

Relational databases will have tables that have relationships or connections to each other. These tables will be linked with primary and foreign keys. A **primary key** is a column that represents the unique identifier for that specific table. Characteristics of primary keys include the following:

- **Uniqueness**: There can be no duplicate primary keys, and each record will have its own distinct key
- **Consistency**: Primary keys will normally remain stable and not change over time as new data is added or deleted
- **Non-nullable**: Primary keys cannot be `null`

A **foreign key** is a column that refers to a primary key from another table. Tables are linked in SQL databases by joining primary and foreign keys. One of the main differences between foreign and primary keys is that foreign keys can contain null values. When a row has a missing foreign key, it indicates that there is no corresponding value in the other table.

The following are common applications and scenarios where data engineers would utilize a relational database:

- **Transactional processing**: Where data from banking, inventory, or commerce is processed.
- **Business intelligence (BI) and business analytics (BA)**: Where data needs to be analyzed and aggregated to extract insights to support business decisions.
- **Customer relationship management (CRM)**: Where customer data needs to be collected and stored. This can include contact information, demographics, and order history.

Advantages of relational databases include the following:

- **Atomicity, consistency, isolation, and durability (ACID) compliance**: Relational databases ensure ACID compliance

- **Security**: Built-in measures such as encryption, access control, and user authentication

- **Data integrity**: Using primary and foreign keys, relational databases ensure accurate and consistent data

- **Normalization**: Minimize data redundancy

Disadvantages of relational databases include the following:

- **Performance**: Slow performance can occur due to complex queries and joins

- **Inflexibility**: Because a schema needs to be established for data storage, limitations are set on the type of data that can be stored

Now that we've reviewed relational databases, we will move on to NoSQL databases in the next section. NoSQL will be important to understand if your team deals with unstructured data.

NoSQL databases

Unlike relational databases, NoSQL databases do not have a relational data model. NoSQL databases are designed to handle unstructured and semi-structured data such as media, audio, and emails. They also don't have a schema, which provides added flexibility in storing and retrieving data.

The following are common applications and scenarios where data engineers would utilize a NoSQL database:

- **Real-time data processing**: Supports real-time analytics and decision-making for streaming data

- **Content management systems (CMSs)**: Where data such as images, documents, or videos need to be stored

- **Search applications**: Where large volumes of unstructured data need to be indexed

Advantages of NoSQL databases include the following:

- **Flexibility**: NoSQL databases can store data in multiple formats, including unstructured, semi-structured, and structured, without a predefined and rigid schema. This is mostly useful when the data requirements are constantly changing.

- **Performance**: NoSQL databases perform much faster while also being able to handle large amounts of data.

Disadvantages of NoSQL databases include the following:

- Not ACID compliant, which makes them an unsuitable option for storing transaction data

- Not suitable for advanced querying

After navigating relational and NoSQL databases, it is time to shift our attention to comprehending the operational details of database systems. We will examine two main categories of data processing systems essential to data engineering in the next section.

OLTP versus OLAP databases

Two popular types of database systems are **online transaction processing (OLTP)** and **online analytical processing (OLAP)** systems.

OLTP databases support day-to-day business transactions and processes where high-speed data entry, updates, and retrieval are a priority. They are designed to handle a high volume of real-time transactions, optimized for simple queries, and are ACID compliant. Typical tasks include updating customer records, inventory management, and processing incoming orders.

OLAP databases are designed to support more complex data analysis. They allow an analyst to perform complex SQL queries and calculations on large volumes of historical data to extract insights where data analysis, query performance, and response time are a priority. Typical tasks include creating forecasts, reports, and other BI activities to influence business decisions.

Normalization

Normalization refers to the process of breaking down large amounts of data into tables and establishing their relationships. This process helps eliminate redundant data through five forms:

1. **First Normal Form (1NF)—eliminate repeating groups**: The most basic form. When this criterion is met, it means the following conditions of a database are satisfied:

    A. Each table has a primary key

    B. All columns contain atomic or indivisible values

    1NF eliminates repeating groups in individual tables while also creating separate tables for each set of related data.

2. **Second Normal Form (2NF)—eliminate redundant data**: The second form of normalization is that all non-primary or foreign key columns are completely dependent on the primary key of a table.

    Each column or attribute must be associated with the data point of that chosen row.

3. **Third Normal Form (3NF)—eliminate columns not dependent on a key**: The third form ensures that all attributes in a table are independent. 3NF provides guardrails against transitive dependency where a value in one column determines the value of another column.

4. **Fourth Normal Form (4NF)—isolate independent multiple relationships**: The fourth form removes multivalued dependencies. Multivalued dependencies occur when one attribute or column determines the values of multiple columns.

5. **Fifth Normal Form (5NF)—isolate semantically related multiple relationships**: The fifth form removes join dependencies. Join dependencies occur when a column from one table determines the values of multiple columns in another table.

Not normalizing a database can cause issues relating to duplicated data, inconsistencies, and accuracy whenever new data is added or manipulated in the database. We have laid a strong foundation for comprehending the fundamental concepts of databases as we conclude this section. We have covered the distinctive features and uses of OLTP and OLAP systems, the landscape of relational databases, and the adaptable world of NoSQL databases.

It's time to expand upon these ideas and explore more challenging areas. We will examine the complexities of databases in the following section.

Must-know advanced database concepts

After going over the basics of databases, it's time to delve into more complicated subjects that are essential tools in the toolbox of a data engineer. We'll summarize various advanced concepts, including triggers, ACID characteristics, and constraints. To guarantee data integrity, consistency, and responsiveness in a database system, each aspect is crucial. Understanding them will improve your conceptual knowledge and your capacity to create and maintain sophisticated database systems.

Remember that this part is not intended to teach you how to do these things from scratch but rather to provide you with a brief overview and prepare you for the kinds of questions and subjects frequently used in interviews for data engineering jobs. Let's clarify things and increase your knowledge of databases.

Constraints

Constraints are rules that are set up to enforce properties in a database. Without constraints, your data will be more prone to inconsistencies and errors. They also ensure referential integrity so that you can reference and retrieve data across multiple tables.

Referential integrity also ensures your data remains synchronized during all modifications to a database. There are several kinds of constraints, such as the following:

- `NOT NULL`: Ensures that a column cannot contain any `NULL` values.

- `UNIQUE`: Ensures that all values in a column are unique.

- `PRIMARY KEY`: Sets a column as the unique identifier for a table. This constraint combines the rules of `NOT NULL` and `UNIQUE`, as a primary key cannot be null and must be unique.

- `FOREIGN KEY`: Ensures that the values in a column will be present in another table.

- `CHECK`: Ensures that values in a column meet a user-defined condition.

- `DEFAULT`: Sets a default value for a column if none is specified.

It's time to look into another critical idea that controls database transactions after knowing the complexities of SQL constraints, how they guarantee the accuracy and dependability of the data, and how they impose specific rules on the data to protect its integrity. ACID is the name of this idea. These characteristics are crucial to any data-driven application's overall robustness and the successful execution of database transactions.

ACID properties

Relational databases support ACID compliance for transactional data. This guarantees the data validity despite mishaps such as power failures or other errors. For a database to be ACID compliant, it must satisfy the following conditions:

- **Atomicity**: The entire record of data will be added to the database, or that transaction will be aborted. There would be no partial entries in the case of a transaction failure or mishap.

- **Consistency**: All data from transactions will follow all predefined business rules and constraints.

- **Isolation**: If transactions occur at the same time, they will not conflict with each other, and each will be executed as normal.

- **Durability**: In the event of any hardware or system failures, all transaction data stored in a database will not be affected or lost.

ACID properties ensure reliable and consistent data that is protected from any system or hardware corruption. Cases where these properties would be implemented include banking, any database storing financial transactions, and customer information.

CAP theorem

In the realm of distributed data systems, the CAP theorem, also known as Brewer's theorem, is a fundamental principle. It asserts that a distributed system cannot simultaneously provide more than two of the three guarantees listed here:

- **Consistency**: Every read returns either the most recent write or an error.

- **Availability**: Every request (read or write) receives a response, but there is no assurance that it contains the most up-to-date version of the data.

- **Partition tolerance**: The system continues to function despite network partitions, which result in lost or delayed communication between nodes.

Simply put, the CAP theorem asserts that a distributed system cannot possess all three of these properties simultaneously. You must choose two.

Here are some examples:

- **Consistency and availability but not partition tolerance**: Traditional relational databases, such as MySQL, Oracle, and Microsoft SQL Server, emphasize consistency and availability. In single-node configurations, where partition tolerance is not an issue (since there is only one node), these systems can guarantee that data is consistent across all transactions and is always accessible for reads and writes. However, network partitions can and do occur in the real world. When these databases are distributed across multiple nodes, they may encounter difficulties in managing partitions, which could necessitate manual intervention or cause downtime.

- **Availability and partition tolerance (AP) but not consistency**: Apache Cassandra is a system that prioritizes availability and partition tolerance. This indicates that even in the event of network partitions, these systems will continue to be accessible, although they may return stale or outdated data. Consider a social media platform where a user's profile picture is updated. Using an availability and partition system may cause some users to temporarily view an outdated image while others view a new one. In this situation, temporary inconsistency is acceptable to maintain the platform's accessibility.

Understanding the CAP theorem is essential for data engineers, as they must frequently make architectural decisions regarding which databases or data systems to employ based on specific requirements and trade-offs of a given application.

The CAP theorem does not prescribe a *one-size-fits-all* solution. Instead, it provides a framework for understanding the trade-offs involved in distributed system design, allowing data engineers to make informed decisions based on the specific needs of their applications.

Triggers

Triggers are sets of code that gets executed when an action occurs. The set of code is normally run after data is manipulated or modified in the database. There are three types of SQL triggers:

- **Data Manipulation Language** (**DML**): Executed when data is inserted, updated, or deleted from a table

- **Data Definition Language** (**DDL**): Executed when the structure of a table is modified, such as the removal or addition of a column

- **Statement**: Executed when certain SQL statements are run in a query

Triggers are used for a variety of tasks, such as enforcing predefined business rules, updating data, running data quality checks, and automating tasks. They can also be used to execute other stored procedures. Because triggers can impact the performance of your database, they should be used sparingly and should not be used for complex actions.

Now that we understand the theoretical foundations of ACID characteristics and how they form the bedrock of dependable and stable data operations, we can put this knowledge to use in the real world. During technical interviews, you will be expected to demonstrate your knowledge by applying these concepts to actual problems. The next step of our trip is answering data engineer interview questions. This material is designed to test your knowledge of the material we've studied so far and to hone your ability to solve problems.

Technical interview questions

The focus of this chapter's final portion shifts to how our database knowledge can be used in technical interviews. Here, we list often-asked interview questions for data engineering positions.

This portion serves two purposes: to assess your comprehension of the fundamental and advanced database concepts presented in prior sections and to improve your ability to communicate your solutions. You can understand the underlying ideas and create a systematic approach to problem-solving because each question is supported with a thorough answer and explanation.

By actively participating in these interview questions, you will strengthen your understanding of database foundations and gain more confidence in your ability to reply to technical questions during job interviews:

- *Question 1*: What is a primary key, and why is it important in a relational database?

  Answer: A primary key is the unique identifier for each row in a table. It cannot be null or non-unique, which enforces data integrity as no two rows will be the same. It also helps establish referential integrity so that relationships can be established between tables.

- *Question 2*: What is database normalization, and why is it important?

  Answer: It is the process of minimizing redundancy and dependencies by enforcing five forms. This helps establish relationships between tables, maintains integrity, and ensures that data is stored consistently.

- *Question 3*: What is the difference between a relational and a non-relational database?

  Answer: A relational database is based on a relational model or schema, where data is organized into tables consisting of rows and columns. This format is best for storing structured data. A non-relational database does not use a schema, and data can be stored without a defined relationship. This format is best for storing semi-structured and unstructured data.

- *Question 4*: What are the ACID properties of a database transaction?

  Answer: Atomicity is where a transaction must either be completed in its entirety or not executed at all. It ensures that partial transactions are not committed. Consistency is when a transaction will follow all defined business rules and database constraints. Isolation occurs when transactions are completely independent from one another. Durability ensures all data is permanent after a transaction is committed, even in the case of system failures.

- *Question 5*: What is a foreign key?

  Answer: A foreign key is a column that references the primary key of another table. This establishes relationships between tables.

Summary

In this chapter, we covered both foundational and advanced database concepts essential for data engineers. We explored different types of databases, normalization, indexing, NoSQL databases, OLTP/OLAP systems, and triggers. Additionally, we provided technical interview questions to test your knowledge and problem-solving skills.

In the next chapter, we will dive into essential SQL for data engineers, focusing on the indispensable SQL skills required for efficient data manipulation and management within databases. Let's continue our journey by mastering essential SQL techniques for data engineering.

8

Essential SQL for Data Engineers

In the world of data engineering, SQL is the unsung hero that empowers us to store, manipulate, transform, and migrate data easily. It is the language that enables data engineers to communicate with databases, extract valuable insights, and shape data to meet their needs. Regardless of the nature of the organization or the data infrastructure in use, a data engineer will invariably need to use SQL for creating, querying, updating, and managing databases. As such, proficiency in SQL can often the difference between a good data engineer and a great one.

Whether you are new to SQL or looking to brush up your skills, this chapter will serve as a comprehensive guide. By the end of this chapter, you will have a solid understanding of SQL as a data engineer and be prepared to showcase your knowledge and skills in an interview setting.

In this chapter, we will cover the following topics:

- Must-know foundational SQL concepts
- Must-know advanced SQL concepts
- Technical interview questions

Must-know foundational SQL concepts

In this section, we will delve into the foundational SQL concepts that form the building blocks of data engineering. Mastering these fundamental concepts is crucial for acing SQL-related interviews and effectively working with databases.

Let's explore the critical foundational SQL concepts every data engineer should be comfortable with, as follows:

- **SQL syntax**: SQL syntax is the set of rules governing how SQL statements should be written. As a data engineer, understanding SQL syntax is fundamental because you'll be writing and reviewing SQL queries regularly. These queries enable you to extract, manipulate, and analyze data stored in relational databases.

- **SQL order of operations**: The order of operations dictates the sequence in which each of the following operators is executed in a query:

  - FROM and JOIN

  - WHERE

  - GROUP BY

  - HAVING

  - SELECT

  - DISTINCT

  - ORDER BY

  - LIMIT/OFFSET

- **Data types**: SQL supports a variety of data types, such as INT, VARCHAR, DATE, and so on. Understanding these types is crucial because they determine the kind of data that can be stored in a column, impacting storage considerations, query performance, and data integrity. As a data engineer, you might also need to convert data types or handle mismatches.

- **SQL operators**: SQL operators are used to perform operations on data. They include arithmetic operators (+, -, *, /), comparison operators (>, <, =, and so on), and logical operators (AND, OR, and NOT). Knowing these operators helps you construct complex queries to solve intricate data-related problems.

- **Data Manipulation Language (DML), Data Definition Language (DDL), and Data Control Language (DCL) commands**: DML commands such as SELECT, INSERT, UPDATE, and DELETE allow you to manipulate data stored in the database. DDL commands such as CREATE, ALTER, and DROP enable you to manage database schemas. DCL commands such as GRANT and REVOKE are used for managing permissions. As a data engineer, you will frequently use these commands to interact with databases.

- **Basic queries**: Writing queries to select, filter, sort, and join data is an essential skill for any data engineer. These operations form the basis of data extraction and manipulation.

- **Aggregation functions**: Functions such as COUNT, SUM, AVG, MAX, MIN, and GROUP BY are used to perform calculations on multiple rows of data. They are essential for generating reports and deriving statistical insights, which are critical aspects of a data engineer's role.

The following section will dive deeper into must-know advanced SQL concepts, exploring advanced techniques to elevate your SQL proficiency. Get ready to level up your SQL game and unlock new possibilities in data engineering!

Must-know advanced SQL concepts

This section will explore advanced SQL concepts that will elevate your data engineering skills to the next level. These concepts will empower you to tackle complex data analysis, perform advanced data transformations, and optimize your SQL queries.

Let's delve into must-know advanced SQL concepts, as follows:

- **Window functions**: These do a calculation on a group of rows that are related to the current row. They are needed for more complex analyses, such as figuring out running totals or moving averages, which are common tasks in data engineering.

- **Subqueries**: Queries nested within other queries. They provide a powerful way to perform complex data extraction, transformation, and analysis, often making your code more efficient and readable.

- **Common Table Expressions (CTEs)**: CTEs can simplify complex queries and make your code more maintainable. They are also essential for recursive queries, which are sometimes necessary for problems involving hierarchical data.

- **Stored procedures and triggers**: Stored procedures help encapsulate frequently performed tasks, improving efficiency and maintainability. Triggers can automate certain operations, improving data integrity. Both are important tools in a data engineer's toolkit.

- **Indexes and optimization**: Indexes speed up query performance by enabling the database to locate data more quickly. Understanding how and when to use indexes is key for a data engineer, as it affects the efficiency and speed of data retrieval.

- **Views**: Views simplify access to data by encapsulating complex queries. They can also enhance security by restricting access to certain columns. As a data engineer, you'll create and manage views to facilitate data access and manipulation.

By mastering these advanced SQL concepts, you will have the tools and knowledge to handle complex data scenarios, optimize your SQL queries, and derive meaningful insights from your datasets. The following section will prepare you for technical interview questions on SQL. We will equip you with example answers and strategies to excel in SQL-related interview discussions. Let's further enhance your SQL expertise and be well prepared for the next phase of your data engineering journey.

Technical interview questions

This section will address technical interview questions specifically focused on SQL for data engineers. These questions will help you demonstrate your SQL proficiency and problem-solving abilities. Let's explore a combination of primary and advanced SQL interview questions and the best methods to approach and answer them, as follows:

- *Question 1*: What is the difference between the WHERE and HAVING clauses?

  Answer: The WHERE clause filters data based on conditions applied to individual rows, while the HAVING clause filters data based on grouped results. Use WHERE for filtering before aggregating data and HAVING for filtering after aggregating data.

- *Question 2*: How do you eliminate duplicate records from a result set?

  Answer: Use the DISTINCT keyword in the SELECT statement to eliminate duplicate records and retrieve unique values from a column or combination of columns.

- *Question 3*: What are primary keys and foreign keys in SQL?

  Answer: A primary key uniquely identifies each record in a table and ensures data integrity. A foreign key establishes a link between two tables, referencing the primary key of another table to enforce referential integrity and maintain relationships.

- *Question 4*: How can you sort data in SQL?

  Answer: Use the ORDER BY clause in a SELECT statement to sort data based on one or more columns. The ASC (**ascending**) keyword sorts data in ascending order, while the DESC (**descending**) keyword sorts it in descending order.

- *Question 5*: Explain the difference between UNION and UNION ALL in SQL.

  Answer: UNION combines and removes duplicate records from the result set, while UNION ALL combines all records without eliminating duplicates. UNION ALL is faster than UNION because it does not involve the duplicate elimination process.

- *Question 6*: Can you explain what a self join is in SQL?

  Answer: A self join is a regular join where a table is joined to itself. This is often useful when the data is related within the same table. To perform a self join, we have to use table aliases to help SQL distinguish the left from the right table.

- *Question 7*: How do you optimize a slow-performing SQL query?

  Answer: Analyze the query execution plan, identify bottlenecks, and consider strategies such as creating appropriate indexes, rewriting the query, or using query optimization techniques such as JOIN order optimization or subquery optimization.

- *Question 8*: What are CTEs, and how do you use them?

  Answer: CTEs are temporarily named result sets that can be referenced within a query. They enhance query readability, simplify complex queries, and enable recursive queries. Use the `WITH` keyword to define CTEs in SQL.

- *Question 9*: Explain the ACID properties in the context of SQL databases.

  Answer: ACID is an acronym that stands for Atomicity, Consistency, Isolation, and Durability. These are basic properties that make sure database operations are reliable and transactional. Atomicity makes sure that a transaction is handled as a single unit, whether it is fully done or not. Consistency makes sure that a transaction moves the database from one valid state to another. Isolation makes sure that transactions that are happening at the same time don't mess with each other. Durability makes sure that once a transaction is committed, its changes are permanent and can survive system failures.

- *Question 10*: How can you handle `NULL` values in SQL?

  Answer: Use the `IS NULL` or `IS NOT NULL` operator to check for `NULL` values. Additionally, you can use the `COALESCE` function to replace `NULL` values with alternative non-null values.

- *Question 11*: What is the purpose of stored procedures and functions in SQL?

  Answer: Stored procedures and functions are reusable pieces of SQL code encapsulating a set of SQL statements. They promote code modularity, improve performance, enhance security, and simplify database maintenance.

- *Question 12*: Explain the difference between a clustered and a non-clustered index.

  Answer: The physical order of the data in a table is set by a clustered index. This means that a table can only have one clustered index. The data rows of a table are stored in the leaf nodes of a clustered index. A non-clustered index, on the other hand, doesn't change the order of the data in the table. After sorting the pointers, it keeps a separate object in a table that points back to the original table rows. There can be more than one non-clustered index for a table.

Prepare for these interview questions by understanding the underlying concepts, practicing SQL queries, and being able to explain your answers.

Summary

This chapter explored the foundational and advanced principles of SQL that empower data engineers to store, manipulate, transform, and migrate data confidently. Understanding these concepts has unlocked the door to seamless data operations, optimized query performance, and insightful data analysis.

SQL is the language that bridges the gap between raw data and valuable insights. With a solid grasp of SQL, you possess the skills to navigate databases, write powerful queries, and design efficient data models. Whether preparing for interviews or tackling real-world data engineering challenges, the knowledge you have gained in this chapter will propel you toward success.

Remember to continue exploring and honing your SQL skills. Stay updated with emerging SQL technologies, best practices, and optimization techniques to stay at the forefront of the ever-evolving data engineering landscape. Embrace the power of SQL as a critical tool in your data engineering arsenal, and let it empower you to unlock the full potential of your data.

In the next chapter, we will dive into the exciting world of data pipeline design for data engineers. Get ready to explore the intricacies of extracting, transforming, and loading data to create robust and efficient data pipelines. Let's continue our quest to become skilled data engineers equipped to tackle any data challenge that comes our way.

Part 3:
Essentials for
Data Engineers Part II

In this part, we will dive deeper into the inner workings of databases, data pipelines, and data warehouses.

This part has the following chapters:

9

Database Design and Optimization

A career in data engineering requires not only handling data but also an understanding of databases, the architectural framework that keeps everything together. Consider attempting to construct a skyscraper without a solid foundation. Eventually, that structure is going to collapse. Any data pipeline can benefit from this as well. No matter how skilled you are with data manipulation and coding, your data pipelines can become unreliable, expensive, and inefficient if you don't have a firm grasp of database design and optimization.

In-depth knowledge of the fundamentals of database design and data modeling will be covered in this chapter, empowering you to create databases that are not only reliable but also scalable and effective. Our aim is to prepare you to become an expert in these vital aspects of data engineering, starting from the fundamentals of tables and relations and progressing to the complexities of indexing and normalization. Whether you are a professional looking to brush up on the basics or a novice entering the field, this chapter tries to give you the conceptual tools necessary to ace any data engineering interview.

The skills you will learn in this chapter are listed as follows:

- Gaining an understanding of the basic ideas behind tables, keys, and relationships—the building blocks of database architecture

- Putting normalization and denormalization strategies into practice to enhance database integrity and performance

- Learning how to evaluate the performance of databases using important metrics such as throughput, latency, and query speed

In this chapter, we will cover the following topics:

- Understanding database design essentials
- Mastering data modeling concepts
- Technical interview questions

Understanding database design essentials

In this section, we will delve into the fundamental principles of database design that are essential for every data engineer. Database design is the process of creating a detailed model of a database. This defines how the data is stored, organized, and manipulated. A well-designed database will be dependent on how the data engineer makes decisions regarding correct data types, constraints, schema design, and **entity-relational (ER)** modeling. This will ensure the data integrity, performance, and reliability of the database.

We will begin by discussing database normalization and the different types:

- **Data normalization**: A procedure that eliminates data duplication and guarantees data integrity. Normalization normally occurs in application databases as opposed to data warehouses. We use normal forms to guide the normalization process. The most common forms include the following:

  - **First Normal Form (1NF)**: Ensures atomicity by organizing data into individual columns, eliminating repeating groups

  - **Second Normal Form (2NF)**: Adds to 1NF by getting rid of partial dependencies and making sure that all non-key attributes depend on the whole primary key

  - **Third Normal Form (3NF)**: Further eliminates transitive dependencies by ensuring that non-key attributes are only dependent on the primary key

- **Other normalization techniques**: In addition to the traditional normalization forms (1NF, 2NF, 3NF), other normalization techniques have emerged to address specific scenarios, such as the following:

  - **Boyce-Codd Normal Form (BCNF)**: A stricter form of normalization that eliminates all dependencies on candidate keys

  - **Fourth Normal Form (4NF)**: Focuses on eliminating multi-valued dependencies within a table

  - **Fifth Normal Form (5NF)**: Deals with eliminating join dependencies and decomposition anomalies

  It is important to understand how to implement normalization to structure your data efficiently and avoid data anomalies.

While normalization is crucial for maintaining data integrity, denormalization is a technique used to optimize database performance in specific situations. By selectively introducing redundancy, denormalization reduces the number of joins required for complex queries, resulting in improved query performance. However, it should be used judiciously, considering the trade-off between performance gains and potential data integrity risks.

- **Normalization and performance trade-offs**: While normalization is crucial for data integrity, it can impact performance due to increased joins and complexity. Understanding the trade-offs between normalization and performance is essential in real-world scenarios. Depending on the specific requirements and use cases, denormalization techniques, such as introducing redundant data or creating summary tables, can be employed to optimize query performance while ensuring data integrity.

- **ER modeling**: ER diagrams (**ERDs**) are graphical representations that visualize the entities, their attributes, and the relationships between them. ERDs utilize symbols such as rectangles (representing entities), lines (representing relationships), and diamonds (representing relationship types) to depict the structure of the data model. By using ERDs, data engineers can easily communicate and visualize complex relationships and dependencies within the database. They consist of the following:

  - **Entities**: Represent real-world objects or concepts, each having a unique identifier known as a primary key

  - **Attributes**: Characteristics or properties of entities that provide additional information

  - **Cardinality or relationships**: Describes the number of occurrences or the participation of entities in a relationship (one-to-one, one-to-many, many-to-many)

  The different relationship types in ER modeling include the following and are visualized in *Figure 9.1*:

  - **One-to-one relationship**: A pattern where each record in one table is associated with only one record in another table

  - **One-to-many relationship**: A pattern where each record in one table is associated with multiple records in another table

  - **Many-to-many relationship**: A pattern where multiple records in one table are associated with multiple records in another table

  The following diagram depicts the different relationships along with their symbols that can appear on a database's ERD:

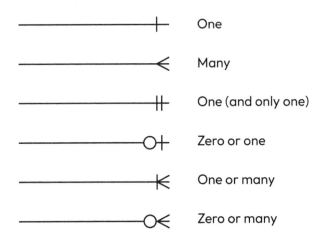

Figure 9.1 – Example of different table relationships in a relational database

Mastering ER modeling allows you to accurately capture the relationships between entities, leading to well-designed databases. The following is a simple example of an ERD for a database of a pizza restaurant. Please note that on the job, you will most likely be working with hundreds of tables:

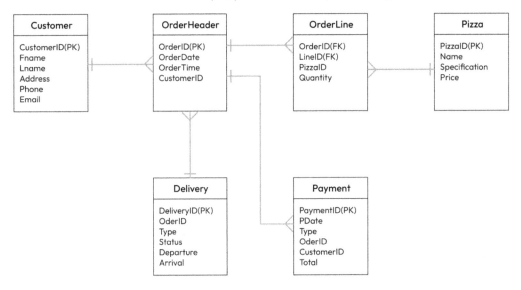

Figure 9.2 – Example of an ERD visualizing the different relationships between the tables

Now that you are aware of the fundamentals of a relational database, we will continue on to database design fundamentals:

- **Schema design**: Schema design involves organizing the structure of the database, including tables, columns, constraints, and relationships. A well-designed schema ensures efficient data retrieval and storage optimization and supports future scalability. Key considerations include the following:

  - **Table structures**: Determine the layout and organization of data within tables, ensuring appropriate column definitions and data types

  - **Indexing**: Enhance query performance by creating indexes on columns frequently used for searching and filtering

  - **Partitioning**: Divide large tables into smaller, manageable pieces based on criteria such as ranges or values, optimizing query performance

  Understanding schema design principles empowers you to create databases that are efficient, flexible, and able to handle increasing data volumes. You will be well equipped to create effective database structures that meet the needs of your organization.

- **Data integrity constraints**: Data integrity constraints are rules applied to a database to help ensure the accuracy and consistency of data. They play an important role in maintaining reliability and integrity. Common constraints include the following:

  - **Primary key**: A unique identifier for each record in a table, ensuring data integrity and enabling efficient data retrieval

  - **Foreign key**: Establishes relationships between tables by referencing the primary key of another table, maintaining data consistency

  - **Check constraint**: Defines specific conditions that must be met for data to be inserted or updated in a column, ensuring data validity

  - **Unique constraint**: Ensures that values in a column or a combination of columns are unique, preventing duplicate entries

  Understanding and implementing these constraints is essential for maintaining data integrity, ensuring data quality, and building reliable systems. Through indexing, they can improve query performance as well.

- **Database design patterns**: Database design patterns are reusable solutions to common database design problems. These patterns focus on how to structure and organize data for manageability and efficiency. Understanding these patterns can guide you in designing databases that align with industry best practices. Examples of database design patterns include the following:

  - **Star schema**: This design pattern is often used in data warehousing and **business intelligence (BI)** systems. In a star schema, one or more *fact tables* reference any number of *dimension tables*, which provide context about the facts. Fact tables represent the central table in a star schema that contains the metrics and measurements of a business process. Their data types will often be numerical. Dimension tables provide context, background information, and attributes for the fact tables. These data types can be text. This schema is useful for simplifying complex database structures and enhancing query performance in **online analytical processing (OLAP)** systems. OLAP systems are software platforms designed specifically for complex data analysis and multidimensional querying. OLAP systems are designed to help analysts and decision-makers explore data in a more interactive and exploratory manner, as opposed to **online transaction processing (OLTP)** systems, which are optimized for routine, day-to-day operations such as data insertion, updating, and deletion. The following diagram visualizes a star schema with fact and dimension tables:

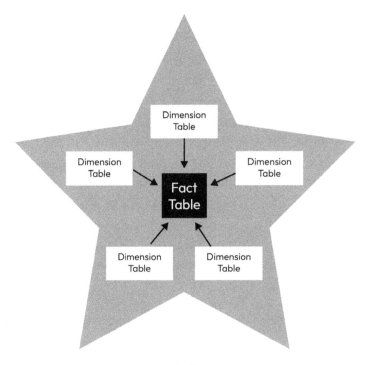

Figure 9.3 – Example of how a star schema is designed

- **Snowflake schema**: This is a variant of a star schema, where the dimension tables are further normalized. While this can save storage space, it can also increase the complexity of SQL queries and decrease overall performance. This is a popular design that can be used to support high volumes of queries and large amounts of data, as we see in the following diagram:

Snowflake Schema

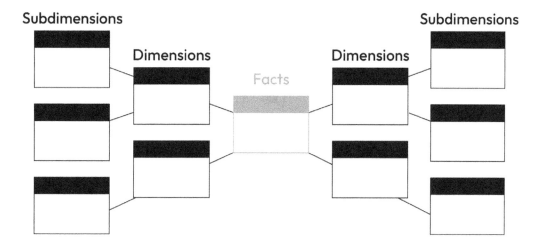

Figure 9.4 – Example of how a snowflake schema is designed

- **Association table pattern**: Also known as a junction table or bridge table pattern, this is primarily used when modeling many-to-many relationships between entities.

- **Hierarchy pattern**: This pattern is used to represent relationships that are *one-to-many* and *tree-structured*—for example, an organizational structure where each employee, except the CEO, has one manager and can have many direct reports.

- **Partitioning pattern**: This pattern is used for large databases to improve performance, manageability, or availability. Data is divided into smaller, more manageable pieces (partitions) based on certain rules (such as date ranges or alphabetically by name).

By utilizing database design patterns, you can create efficient and scalable databases while ensuring data integrity.

- **Database design tools**: In the realm of database design, there are several tools available that can aid in the design and modeling process. Familiarizing yourself with these tools can significantly enhance your efficiency and productivity. Some commonly used database design tools include the following:

  - **ERD tools**: Software such as PowerPoint, Figma, or Canva enables you to create detailed diagrams, visually representing relationships between entities, attributes, and their cardinality.

  - **Database design software**: Tools such as MySQL Workbench, **Microsoft SQL Server Management Studio** (**Microsoft SSMS**), or Oracle SQL Developer provide integrated environments for designing and managing databases. These tools offer graphical interfaces for designing tables, relationships, and constraints, as well as generating SQL scripts for implementation.

By leveraging these tools effectively, you can streamline the database design process and collaborate seamlessly with other stakeholders.

By considering these additional concepts and aspects of database design, including tools, trade-offs, and documentation, you will develop a well-rounded understanding of the subject. This knowledge will enable you to design databases that not only meet functional requirements but also exhibit optimal performance, maintainability, and scalability.

Now that we have covered the fundamentals of database design, we will continue on to indexing strategies in the next section.

Indexing

An index makes it possible for databases to retrieve data more quickly. The database engine would have to search through every row in a table in the absence of an index, which would be analogous to searching through every book in a library to find a particular piece of information. A subset of the dataset is stored in a data structure created by indexing, which is arranged to make it easier for the database engine to locate the necessary data. Although it's important to remember that maintaining an index can slow down write operations because the index itself needs to be updated, the goal is to speed up read operations.

Various types of indexes cater to different needs and database structures:

- **Single-column**: The simplest kind of index is a single-column index, which is created on a single table column.

- **Multi-column**: Often referred to as a composite index, this type of index incorporates two or more table columns.

- **Unique index**: Makes sure each value in the indexed column is distinct by using a unique index. Usually, this is used for the main key columns.

- **Full-text**: With a focus on text-based columns, this kind of index is best suited for word searches inside textual data.

- **Clustered**: Data retrieval is accelerated by clustered indexes, which store data in tables in the same physical order as the index.

- **Non-clustered**: In this case, the rows' physical storage order on disk is not consistent with the logical order of the index.

Here are some indexing best strategies:

- **Selectivity**: An index performs better when the data in a column is more distinctive. Optimal selectivity is preferred.

- **Keep it light**: Having too many indexes can cause write operations to lag. Select indexes carefully.

- **Maintenance**: To keep performance at its best, periodically rebuild indexes and update statistics.

Consider a data engineer for a major e-commerce platform who is tasked with managing a customer database. Millions of rows, each representing a transaction, make up the Orders table, which is regularly queried to produce reports. Getting every order for a particular customer is a common query.

A time-consuming and resource-intensive process, searching through every row in the Orders table, would be required for each query if there were no index on the `customer_id` column. The data engineer could improve both the efficiency of the data pipeline as a whole and the generation of individual reports by implementing a single-column index on `customer_id` and significantly cutting down on query time.

As we've seen, indexing is a key tactic for enhancing database performance, particularly for operations that involve a lot of reading. Good indexing can make the difference between a user-friendly application that loads quickly and one that takes a long time.

Data partitioning

As database size and complexity increase, so do the difficulties associated with managing them. Large datasets become more manageable, scalable, and performant as a result of data partitioning as a solution to these problems. This section will dissect the two primary types of data partitioning—horizontal and vertical—and provide implementation best practices for each. Understanding data partitioning is crucial whether you're working with a monolithic database that requires more efficient query performance or a distributed system that must scale horizontally. Let's examine techniques and considerations that make data partitioning a pillar of robust database design.

Horizontal partitioning, also known as sharding, is the process of dividing a table into smaller tables, each of which contains a subset of the data. These smaller tables are referred to as shards. Shards may reside on the same server or multiple servers. The objective is to partition the dataset so that each shard is more manageable and can be independently accessed or updated. This strategy is especially effective for enhancing read and write speeds of large, distributed databases.

As an alternative to dividing the table into smaller tables horizontally, **vertical partitioning** divides the table into smaller tables vertically. Instead of rows, in this instance, each partition has a subset of the columns. To separate frequently accessed columns from infrequently accessed ones, vertical partitioning is frequently utilized. This can lead to increased disk I/O efficiency and shortened query times.

Here are some data partitioning best practices:

- **Even data distribution**: Aim for an even distribution of data across all partitions, whether horizontal or vertical, to prevent any one partition from becoming a bottleneck.

- **Partitioning key**: In horizontal partitioning, it is crucial to select an appropriate partitioning key. The key should be selected to ensure uniform data distribution.

- **Regular monitoring**: Over time, data growth or reduction can cause partition size imbalances, necessitating partition rebalancing.

Consider a scenario in which a data engineer is responsible for managing a real-time analytics platform that processes large volumes of sensor data for industrial machinery. The data is primarily stored in `sensor_id` and `sensor_value` columns. As data inflow rates grow, query performance begins to degrade.

To address this issue, the data engineer may employ horizontal partitioning, distributing data based on `sensor_id`. This would restrict each shard to a particular range of sensor IDs, thereby enhancing the performance of sensor-specific queries.

For columns that are rarely accessed, such as metadata columns, the engineer may use vertical partitioning to separate these columns into a separate table. This would optimize disk I/O, as the majority of queries would only need to scan frequently accessed columns.

Understanding the complexities of data partitioning as a data engineer is crucial for scaling database systems and ensuring optimal performance.

Performance metrics

In any database system, performance is not a luxury but an absolute requirement. Inadequate performance can result in operational delays, dissatisfied users, and missed opportunities. This section focuses on vital database performance metrics such as query speed and latency, which serve as the environment's pulse. Understanding these metrics is essential for identifying problems, making well-informed decisions, and optimizing the system for current and future demands. In addition, we will examine the tools that can assist you in keeping a close eye on these metrics, enabling proactive management of your database system. Let's examine the metrics that make or break your database's performance.

Query speed is a crucial metric that influences the user experience and system efficiency directly. Users are able to quickly retrieve the information they require, resulting in improved decision-making and streamlined operations. Frequently, you will need to optimize database schemas, improve indexing, and fine-tune SQL queries to increase query speed. Keep a close eye on any frequently slow-performing queries, as they may become system bottlenecks.

Latency is the time it takes to process a request from the time it is made until a response begins to arrive. Low database latency facilitates faster data retrieval, which is essential for real-time applications and services. Network latency and disk I/O are two variables that can influence database latency. Minimizing latency often involves optimizing your infrastructure and potentially distributing data geographically closer to where it is most often accessed.

Monitoring tools are crucial for keeping track of these metrics. Tools such as Prometheus, Grafana, and built-in database monitoring systems can provide real-time visibility into query performance, latency, and other vital statistics. These tools can detect issues before they become critical, enabling proactive troubleshooting and performance enhancement.

Performance metrics serve as a yardstick for measuring the health and efficacy of a database system. Without an understanding of these metrics, indexing and partitioning optimizations cannot be effectively validated.

Designing for scalability

In today's ever-changing data landscape, the ability to adapt and expand is crucial. Scalability is not merely a buzzword; it is a fundamental aspect of modern database design that ensures your system can efficiently manage increasing workloads. This section explores the fundamentals of designing for scalability, including horizontal and vertical scaling options, robust replication strategies, and effective load balancing. Regardless of whether your database serves a start-up or a multinational corporation, a well-planned scalability strategy is essential for meeting current and future demands. Let's examine the fundamentals that will allow your database to expand with your business.

Scalability in database systems typically boils down to two primary options: horizontal scaling and vertical scaling. Vertical scaling is the process of adding more resources to an existing server, such as CPU, RAM, and storage. Although this is a simple method, it has limitations, particularly in terms of hardware capabilities and expense. Horizontal scaling, on the other hand, involves adding more servers to distribute the load. This method is more adaptable and, in theory, infinitely scalable, but it introduces complications regarding data distribution and consistency.

Replication strategies: Replication is another important factor to consider when designing for scalability. It involves duplicating your database for **high availability** (**HA**) and **fault tolerance** (FT). Consider the following replication strategies:

- **Master-slave replication**: In this model, all write operations are directed to the master server, while any of the slave servers can process read operations

- **Master-master replication**: In this configuration, any server can perform both read and write operations, making the system more robust but more difficult to manage due to consistency concerns

- **Sharding with replication**: Combining sharding (horizontal partitioning) with replication can provide both scalability and reliability, but it requires careful planning and management

Load balancing is the process of distributing incoming database queries across multiple servers so that no single server becomes overloaded. This is particularly important for horizontally scalable architectures. Utilizing algorithms such as round-robin or least connections, load balancers can be configured to efficiently route queries to the least occupied resources.

Mastering data modeling concepts

In this section, we will explore in detail essential data modeling concepts that every data engineer must be familiar with. Data modeling is the process of structuring and organizing data to represent real-world entities, their attributes, and the relationships between them. A solid understanding of data modeling concepts is crucial for designing efficient and accurate databases.

The following diagram depicts three different data models that can be found in a database:

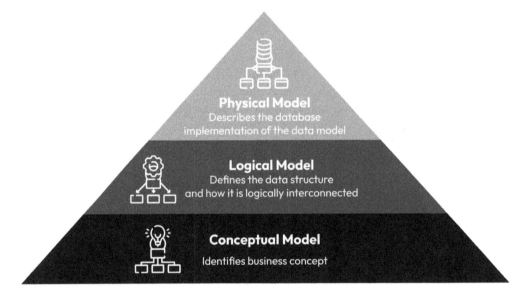

Figure 9.5 – Three different types of data models

Let's look at these in more detail:

- **Conceptual data model**: The conceptual data model represents high-level business concepts and relationships without concerning itself with implementation details. It focuses on capturing essential entities, their attributes, and the associations between them. The conceptual data model serves as a communication tool between data engineers, stakeholders, and domain experts to ensure a shared understanding of the business requirements.

- **Logical data model**: The logical data model provides a more detailed representation of the data, refining the conceptual model into a structure that is closer to implementation. It involves translating the conceptual model into a set of entities, attributes, and relationships. The logical data model aims to capture the semantics and meaning of the data in a technology-agnostic manner. Common techniques used in logical data modeling include ERDs and **Unified Modeling Language (UML)** diagrams.

- **Physical data model**: The physical data model focuses on the actual implementation of the database design, considering the specific **database management system (DBMS)** and its features. It defines tables, columns, data types, constraints, indexes, and other implementation details. The physical data model aims to optimize performance, storage efficiency, and data integrity based on the requirements and constraints of the target DBMS.

By mastering these data modeling concepts, including conceptual, logical, and physical data modeling, as well as ERDs, normalization, advanced modeling techniques, and data modeling tools, data engineers can create well-structured, efficient, and scalable databases that align with business requirements.

Now that we have covered database design fundamentals, we will test your understanding with sample interview questions.

Technical interview questions

Having covered essential concepts in database design and optimization, we will now provide a series of questions to help you effectively prepare for your interview. We have also provided an example answer for each question:

- *Question 1*: What is the purpose of data normalization, and what are the different normal forms?

  Answer: Data normalization is the process of organizing data in a database to minimize redundancy and improve data integrity. The different normal forms are as follows:

  - **First Normal Form (1NF)**: Ensures atomicity by eliminating repeating groups and storing each data value in a separate column

  - **Second Normal Form (2NF)**: Builds upon 1NF by eliminating partial dependencies, ensuring that non-key attributes are fully dependent on the primary key

  - **Third Normal Form (3NF)**: Further eliminates transitive dependencies, ensuring that non-key attributes are only dependent on the primary key and not on other non-key attributes

  - **Fourth Normal Form (4NF)**: Focuses on eliminating multi-valued dependencies within a table

  - **Fifth Normal Form (5NF)**: Deals with eliminating join dependencies and decomposition anomalies

Further explanation: Data normalization is crucial for maintaining data integrity and minimizing redundancy in a database. 1NF ensures that each data value is stored in a separate column, eliminating repeating groups. 2NF eliminates partial dependencies, ensuring non-key attributes depend on the entire primary key. 3NF removes transitive dependencies, ensuring that non-key attributes only depend on the primary key. By applying normalization, we create well-structured databases that minimize data duplication and maintain data consistency.

- *Question 2*: Explain the difference between a primary key and a foreign key.

  Answer: A primary key is an identifier that is unique to each record in a table. It ensures data integrity and provides a means to identify and retrieve records in a unique manner. In contrast, a foreign key establishes a connection between two tables by referencing the primary key of another table. It maintains consistency and enforces referential integrity between related tables.

  Further explanation: A primary key is a column or a set of columns that uniquely identifies each record in a table. It ensures data integrity and provides a way to retrieve records efficiently. A foreign key, on the other hand, is a column or a set of columns in a table that references the primary key of another table. It establishes a relationship between the two tables and enforces referential integrity by ensuring that the foreign key values correspond to existing primary key values in the referenced table. This maintains data consistency and allows us to create meaningful associations between related data.

- *Question 3*: How do you optimize database performance?

  Answer: Optimizing database performance involves various techniques, including the following:

  - **Indexing**: Creating indexes on frequently queried columns to improve search and retrieval speed

  - **Query optimization**: Analyzing and optimizing SQL queries, including proper indexing, efficient join operations, and query restructuring

  - **Denormalization**: Introducing controlled redundancy by selectively combining related data into a single table or duplicating data for performance improvements

  - **Partitioning**: Dividing large tables into smaller, manageable partitions based on predefined criteria such as ranges or values to enhance query performance

  - **Caching**: Implementing caching mechanisms to store frequently accessed data in memory for faster retrieval

Further explanation: To optimize database performance, we can employ several techniques. Indexing plays a crucial role by creating indexes on frequently queried columns, allowing for faster search and retrieval. Query optimization involves analyzing and optimizing SQL queries, ensuring appropriate indexing, efficient join operations, and query restructuring. Additionally, denormalization can be used to introduce controlled redundancy by combining related data or duplicating data to reduce the need for complex joins. Partitioning large tables into smaller partitions based on specific criteria can also enhance query performance. Lastly, implementing caching mechanisms, such as in-memory caching, can significantly improve performance by storing frequently accessed data for faster retrieval.

Remember to tailor your answers to your own experiences and knowledge, using these examples as a guide to structure your responses effectively.

Summary

In this chapter, we explored vital aspects of database design and optimization. We delved into key concepts of database design, including normalization, ER modeling, and schema design. We also discussed advanced topics such as denormalization, data integrity constraints, and performance optimization techniques.

By mastering these concepts, you have developed a strong foundation in database design, which is crucial for excelling in data engineering roles. Additionally, you have prepared yourself to confidently answer interview questions, demonstrating your knowledge and practical experience in designing efficient and scalable databases.

In the next chapter, we will delve into the exciting world of data processing and transformation. We will explore techniques, tools, and best practices for extracting, transforming, and loading data to enable meaningful analysis and insights. Get ready to unlock the power of data transformation and manipulation!

10

Data Processing and ETL

Navigating the intricacies of data engineering roles requires an in-depth understanding of data processing and **Extract, Transform, and Load** (**ETL**) processes. Not only do these foundational skills form the foundation upon which data pipelines are constructed but they are also integral components of the data engineering interview landscape. Therefore, mastering them is a prerequisite for anyone seeking success in data engineering roles. In this chapter, we will delve into the nitty-gritty details of implementing ETL processes, examine the various paradigms of data processing, and guide you on how to prepare for technical data engineering interview questions. This chapter aims to equip you with the knowledge and skills necessary to ace data engineering interviews by providing real-world scenarios, technical questions, and best practices.

In this chapter, we will cover the following topics:

- Fundamental concepts
- Practical application of data processing and ETL
- Preparing for technical interviews

Fundamental concepts

Before delving into the complexities of ETL and data processing, it is essential to lay a solid foundation by understanding the underlying concepts and architectures. This section serves as a guide to the fundamental concepts that every data engineer should understand. By the end of this section, you should have a comprehensive understanding of the essential frameworks and terminology for both practical applications and interview success.

The life cycle of an ETL job

The life cycle of an ETL job is a well-orchestrated sequence of steps designed to move data from its source to a destination, usually a data warehouse, while transforming it into a usable format. The process begins with extraction, the phase in which data is extracted from multiple source systems. These systems could be databases, flat files, application programming interfaces, or even web scraping targets. The key is to extract the data in a manner that minimizes the impact on source systems, which is typically accomplished using techniques such as incremental loads or scheduled extractions during off-peak times.

Transformation, often considered the core of the ETL process, is the next phase. During this phase, the raw data is transformed into a format that is suitable for analysis or reporting. Transformations may include data cleansing, aggregation, enrichment with additional information, and reorganization. Most data quality checks are performed at this stage, ensuring that the data is accurate, consistent, and usable.

Then, we finally reach the loading phase. The transformed data is loaded into the target system, which is typically a data warehouse. The loading procedure must be effective and conform to the requirements of the target system. Some data warehouses, for instance, favor bulk inserts for performance reasons, whereas others may support more complex operations. During the loading phase, data indexing and partitioning occur to optimize the data for quick retrieval and analysis.

Understanding the life cycle of an ETL job is essential, as each phase has its own set of best practices, obstacles, and tooling alternatives. Mastering these facets will not only equip you to design and implement ETL pipelines proficiently but also confidently answer related interview questions.

Practical application of data processing and ETL

The next logical step, after mastering the fundamental concepts and architectures, is to apply this knowledge to real-world scenarios. This section focuses on the practical aspects of ETL and data processing, guiding you through the entire pipeline creation process—from design to implementation and optimization. Whether you're constructing a simple data ingestion task or a complex, multi-stage ETL pipeline, the hands-on exercises and case studies in this section will equip you with the knowledge and confidence to overcome any data engineering challenge. By the end of this section, you will not only be equipped to implement effective ETL solutions but also to excel in interview questions pertaining to this topic.

Designing an ETL pipeline

Designing an ETL pipeline is the crucial first procedure that sets the stage for the implementation and optimization phases that follow. The initial step in this procedure is requirements collection. Understanding the business requirements and technical constraints is essential for defining the ETL pipeline's scope and objectives. This requires discussions with stakeholders, such as data analysts, business leaders, and other engineers, to determine what data must be moved, how often, and what transformations are required.

Following the clarification of requirements, the next step is **schema design**. This involves defining the data's structure in the target system, which is typically a data warehouse. During this phase, decisions are made regarding data types, primary and foreign keys, and indexing strategies. Schema design is not merely a technical task; it requires an in-depth knowledge of the business domain to ensure that the data model supports the types of queries and analyses that end users will execute.

The third essential factor is **data mapping**. Here, you specify how source system data maps to the target schema. Complex data mapping may involve multiple transformations, aggregations, or calculations. This step typically entails the creation of a detailed mapping document that will serve as the blueprint for the implementation phase.

Designing an ETL pipeline is a delicate balancing act requiring both technical and business acumen. A pipeline that is well designed will not only meet the current needs but will also be adaptable enough to accommodate future needs. It sets the foundation for a robust, scalable, and efficient data movement process, making it an essential skill for any data engineer.

In this section, we will explore the essential concepts of data processing for data engineers. Data processing involves the manipulation, cleansing, and transformation of data to derive meaningful insights and facilitate decision-making.

Implementing an ETL pipeline

Implementing your ETL pipeline is the next crucial step following the completion of the design phase. The initial activity in this phase is tool selection. The selection of ETL tools and technologies can significantly affect a pipeline's efficiency and maintainability. Whether you choose traditional ETL tools such as Informatica, cloud-based solutions such as AWS Glue, or programming languages such as Python, your selection should be in line with the project's requirements, scalability requirements, and your team's expertise.

Following the selection of the proper tools, the focus shifts to coding best practices. Implementing an ETL pipeline requires more than just writing code to move and transform data; it also requires doing so in a clean, efficient, and maintainable manner. This involves following coding standards, commenting, and documentation. Modularizing your code to make it reusable and easier to test is also essential.

Error handling is another crucial aspect of ETL implementation, related to testing. During data extraction, transformation, or loading, robust error-handling mechanisms must be in place to identify and log any potential errors. This ensures that the pipeline is resistant to failures and that problems can be quickly identified and resolved.

Implementation is where strategy meets reality. In this phase, your meticulously crafted plans and designs become operational. Not only is a well-implemented ETL pipeline functional but it is also robust, scalable, and simple to maintain. This ability is essential for data engineers and is frequently emphasized in technical interviews.

Optimizing an ETL pipeline

Once your ETL pipeline is operational, you should shift your focus to optimization. The first consideration is performance tuning. Even a properly designed and implemented pipeline may experience performance bottlenecks. These may result from inefficient queries, sluggish data transformations, or inefficient loading strategies. Identifying and resolving these bottlenecks typically requires profiling the pipeline, analyzing logs, and implementing targeted code or configuration improvements.

Next, we concentrate on scalability. Your ETL pipeline must be capable of scaling to accommodate growing data volumes and changing business requirements. Often, scalability considerations involve selecting the appropriate hardware, optimizing data partitioning, and possibly transitioning to a distributed computing environment. The objective is to construct a pipeline that not only meets current demands but is also adaptable enough to meet future challenges.

Monitoring and logging come last, but not least. A trustworthy ETL pipeline is optimized and trust is derived from visibility. Implementing extensive monitoring and logging enables you to monitor the pipeline's health, receive alerts on failures or performance issues, and gain insights for further optimization. This ensures the pipeline's durability and maintainability over time.

The optimization process is ongoing. It requires a keen eye for detail, a comprehensive understanding of the available data and tools, and a dedication to continuous improvement. Not only will mastering this aspect of ETL make your pipelines more robust and efficient but it will also give you a substantial advantage in technical interviews.

Now that we have covered the ETL process, we will apply this knowledge to example interview questions.

Preparing for technical interviews

In this section, we will prepare you for technical interview questions specifically focused on ETL and data processing. These questions aim to assess your understanding of the concepts and practical considerations involved in ETL workflows and data processing.

To excel in the technical interview, focus on the following areas:

- *Data transformation techniques*: Be prepared to discuss different data transformation techniques, such as data aggregation, normalization, denormalization, and feature engineering. Provide examples of how you have applied these techniques in real-world scenarios and the benefits they brought to data analysis and decision-making processes.

- *ETL best practices*: Demonstrate your knowledge of the ETL best practices, including data quality checks, error handling mechanisms, and data validation techniques. Explain how you ensure data accuracy, completeness, and consistency during the ETL process. Showcase your experience in dealing with large datasets and explain how you handle scalability and performance challenges.

- *Data pipeline optimization*: Discuss strategies for optimizing data-processing pipelines. Highlight techniques such as parallel processing, caching, and query optimization to improve performance and reduce latency. Provide examples of how you have optimized data pipelines to handle large volumes of data efficiently.

- *Data integration challenges*: Be prepared to discuss the challenges involved in integrating data from diverse sources. Address issues such as data format compatibility, schema mapping, data deduplication, and handling data inconsistencies. Illustrate your problem-solving skills by providing examples of how you have tackled data integration challenges in your previous projects.

- *Real-time data processing*: Demonstrate your understanding of real-time data processing and its importance in today's fast-paced environments. Explain the differences between batch processing and real-time streaming, and discuss the technologies or frameworks you have worked with, such as Apache Kafka, Apache Flink, or Spark Streaming.

Now, let's dive into five sample technical interview questions related to ETL and data processing, along with example answers:

- *Question 1*: How would you handle missing or erroneous data during the ETL process?

  Answer: When encountering missing or erroneous data, I would implement data quality checks to identify and handle these issues. For missing data, I would assess the impact on downstream analysis and determine the appropriate approach, such as imputation or exclusion. For erroneous data, I would apply validation rules and data cleansing techniques, including outlier detection, data type validation, and range checks. By incorporating error handling mechanisms, such as logging and alerting, I can ensure that data anomalies are detected and resolved promptly.

- *Question 2*: How do you optimize a data-processing pipeline for better performance?

  Answer: To optimize a data-processing pipeline, I focus on several aspects. Firstly, I parallelize tasks and leverage distributed computing frameworks such as Apache Spark to enable efficient processing across multiple nodes. Secondly, I optimize data access by implementing caching mechanisms and using appropriate data structures for faster retrieval. Additionally, I employ query optimization techniques, such as indexing and partitioning, to reduce the data volume processed. Lastly, I continuously monitor the pipeline's performance, analyze bottlenecks, and fine-tune the system based on the observed patterns and workload characteristics.

- *Question 3*: What are some key considerations when integrating data from multiple sources?

  Answer: When integrating data from multiple sources, it is crucial to consider data format compatibility, data schema mapping, and data consistency. I ensure that data formats are aligned or appropriately transformed to enable seamless integration. Schema mapping involves mapping attributes from different sources to a unified schema, resolving conflicts, and handling data transformations if required. Data deduplication techniques, such as record matching and merging, ensure data consistency and eliminate redundant information. By addressing these considerations, I establish a cohesive view of the integrated data, ensuring its reliability and usability across the organization.

- *Question 4*: How would you handle the processing of large volumes of data in a data pipeline?

  Answer: Processing large volumes of data requires careful planning and optimization. Firstly, I would leverage distributed processing frameworks such as Apache Hadoop or Apache Spark to enable parallel processing across multiple nodes, taking advantage of their scalability and fault tolerance. Additionally, I would implement techniques such as data partitioning, where data is divided into smaller chunks based on specific criteria, allowing for efficient processing. I would also employ data compression techniques to reduce storage requirements and optimize network transfer. By optimizing resource utilization, adopting efficient algorithms, and leveraging distributed computing capabilities, I ensure that the data pipeline can handle the processing of large volumes of data effectively.

- *Question 5*: How does real-time data processing differ from batch processing?

  Answer: Real-time data processing and batch processing differ in their timing and processing approach. Batch processing involves processing data in large volumes at specific intervals, often during off-peak hours. It is well suited to processing historical or accumulated data where immediate analysis is not required. On the other hand, real-time data processing involves the ingestion, processing, and analysis of data as it arrives, enabling real-time insights and immediate actions. Real-time processing is ideal for scenarios in which quick decision-making or immediate reactions to data events are necessary. Technologies such as Apache Kafka or Spark Streaming facilitate real-time data processing by enabling the continuous ingestion and processing of data streams.

Summary

In conclusion, by familiarizing yourself with these technical interview questions and their example answers, you will be better prepared to showcase your knowledge and expertise in ETL and data processing. Remember to tailor your responses to your own experiences and projects, providing concrete examples to demonstrate your practical understanding of these concepts.

Our next chapter will cover data pipeline design. Best of luck in your data engineering journey and interviews!

11
Data Pipeline Design for Data Engineers

Understanding databases, **Extract**, **Transform**, **Load** (ETL) procedures, and data warehousing is only the beginning of negotiating the tricky terrain of data engineering interviews. You also need to be an expert at designing and managing data pipelines. A well-designed data pipeline is the lifeblood of any data-driven organization, regardless of whether you are processing real-time data streams or orchestrating large-scale batch processes. This chapter aims to be your in-depth reference on this important topic, tailored to give you the information and abilities you need to ace the interview. We'll examine the underlying principles of data pipeline architecture, go over how to create a successful data pipeline, and then put your knowledge to the test with real-world technical interview questions.

In this chapter, we will cover the following topics:

- Data pipeline foundations
- Steps to design your data pipeline
- Technical interview questions

Data pipeline foundations

A data pipeline is a set of processes and technologies designed to transport, transform, and store data from one or more sources to a destination. The overarching objective is frequently to facilitate the collection and analysis of data, thereby enabling organizations to derive actionable insights. Consider a data pipeline to be similar to a conveyor belt in a factory: raw materials (in this case, data) are taken from the source, undergo various stages of processing, and then arrive at their final destination in a refined state.

The following diagram depicts the typical stages of a data pipeline:

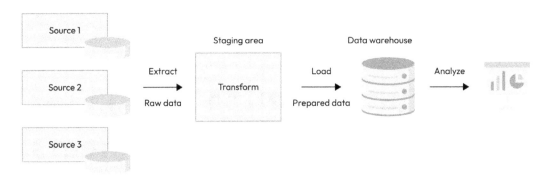

Figure 11.1 – Example of a typical data pipeline

A typical data pipeline comprises four primary components:

- **Data sources**: These are the origins of your data. Sources of data include databases, data lakes, APIs, and IoT devices.

- **Data processing units** (**DPUs**): DPUs are the *factory floor* where raw data is transformed. They involve steps such as cleaning, enriching, and aggregating data to make it more comprehensive.

- **Data sink**: A location where processed data is deposited. It could be a data warehouse, another database, or a visualization dashboard.

- **Orchestration**: This is the *conductor* of the pipeline, managing the flow of data from source to sink. It schedules tasks, handles errors, and ensures that data flows without interruption.

Any aspiring data engineer must have a thorough understanding of data pipelines. These pipelines serve as the foundation of modern data architectures, allowing businesses to make decisions based on data. They can scale to handle large volumes of data, adapt to a variety of data formats, and guarantee that data is accurate, consistent, and readily accessible when required.

Now that we have a fundamental understanding of what a data pipeline is, let's examine the various types of data pipelines that you may encounter, including batch and stream processing, as well as ETL and **Extract**, **Load**, **and Transform** (**ELT**) paradigms.

Types of data pipelines

Each type and flavor of data pipeline has its own benefits and disadvantages. In general, they can be classified according to the processing method (batch versus stream) and the data flow pattern (ETL versus ELT). Let's have a look at this in more detail:

- **Batch processing**: In batch processing, data is gathered over a specific time frame and then processed in bulk. This method is efficient and cost-effective for non-time-sensitive operations. Nonetheless, waiting for a batch to be processed can cause some delays—for example, a retailer gathers sales information throughout the day and runs a nightly batch process to update their inventory and generate reports.

- **Stream processing**: Stream processing, on the other hand, processes data in real time or near real time, as it arrives. This is beneficial for applications that require immediate action based on incoming data. For example, a bank's fraud detection system analyzes transactions in real time to immediately flag suspicious activities.

- **ETL pipelines**: In ETL pipelines, data is extracted from the source, converted to the desired format, and then loaded into the destination database or data warehouse. For example, a healthcare organization collects patient records from multiple clinics, standardizes the data, and then loads it into a centralized data warehouse for analytics.

- **ELT**: In ELT, the data is initially extracted and loaded into the destination, followed by transformation. This strategy is popular in transformation-efficient data warehouses such as Snowflake and BigQuery. For example, an e-commerce platform streams clickstream data into a data lake and then transforms and analyzes the data using a cloud-based data warehouse.

Having discussed the different types of data pipelines, it's crucial to understand the key components that make them operational. Let's proceed to dissect these building blocks for a more rounded grasp of data pipeline architectures.

Key components of a data pipeline

Each component of a data pipeline serves a specific purpose in order for the system as a whole to function optimally. Understanding these essential components is crucial not only for designing robust pipelines but also for discussing them in interviews. In this section, we will examine the four pillars – data sources, DPUs, data sinks, and orchestration:

- **Data sources**: Data sources are the origination points of data within a data pipeline. These can include structured databases such as SQL databases, semi-structured sources such as JSON files, and unstructured data such as images or logs. For example, a platform for social media may collect user interaction data from its mobile app (JSON), server logs (text files), and relational databases containing user profiles.

- **DPUs**: DPUs serve as the pipeline's transformation engine. The data is cleansed, enriched, aggregated, and otherwise prepared for its final destination. Typically, technologies such as Apache Spark and Hadoop are employed for this purpose. For example, a weather forecasting system may collect raw data from multiple sensors, eliminate noisy measurements, aggregate data points, and employ **machine learning** (**ML**) algorithms to predict weather conditions.

- **Data sinks**: Data sinks are the storage or display locations for processed data. This could be a data warehouse (such as Snowflake), a NoSQL database (such as MongoDB), or a real-time dashboard. For example, an IoT company that monitors industrial machines could send processed sensor data to a real-time dashboard for operational teams and store it in a data warehouse for long-term analytics.

- **Orchestration**: The control tower of a data pipeline, orchestration is responsible for managing the flow of data from source to sink. It schedules tasks, manages errors, and guarantees data integrity. For example, orchestration may be used by an online streaming service to ensure that user viewing history, which is stored in a database, is routinely processed to update recommendation algorithms without causing system outages or data inconsistencies.

Now that we have a good grasp of the key components that make up a data pipeline, it's time to explore the step-by-step process of designing one. This will further your understanding and help you answer design questions in interviews effectively.

Steps to design your data pipeline

Similar to building a structure, designing a data pipeline requires careful planning, a solid foundation, and the proper tools and materials. In the realm of data engineering, the blueprint represents your design process. This section will guide you through the essential steps involved in designing a reliable and efficient data pipeline, from gathering requirements to monitoring and maintenance:

1. **Requirement gathering**: The initial step in designing a data pipeline is to comprehend what you are building and why. Collect business and data requirements to comprehend the project's scope, objectives, and limitations. For example, to increase sales, an online retailer wants to analyze customer behavior. The data requirements may specify the use of real-time analytics, while the business requirements may include the monitoring of customer interactions.

2. **Identify data sources**: Once you have determined what you require, determine where to obtain it. This involves identifying the data source types and ensuring that you can access and extract data from them. For example, researchers may need to collect data from hospital records, government databases, and wearable health devices for a public health study.

3. **Data processing and transformation**: After identifying the data sources, determine how the data will be cleansed, transformed, and enriched. Based on your needs, choose the appropriate DPUs and technologies. For example, a financial institution may need to collect transaction data from multiple banks, cleanse it by removing duplicates, and then convert it to a format that's suitable for risk analysis.

4. **Data storage**: The next step is to determine the storage location for the processed data. This could be a data lake, a data warehouse, or another storage solution, depending on your needs. For example, a media organization may store processed viewer metrics in a data warehouse for historical analysis while employing a NoSQL database for real-time analytics.

5. **Data orchestration**: Orchestration involves controlling the flow of data through your pipeline. Choose an orchestration tool capable of handling the complexity of your pipeline and ensuring that data flows smoothly from source to sink. For example, Apache Airflow may be utilized by an e-commerce platform to schedule daily batch jobs that update product recommendations based on user behavior.

6. **Monitoring and maintenance**: Plan how you will monitor the pipeline and handle failures or inconsistencies as a final step. Define **key performance indicators (KPIs)** and establish mechanisms for alerting and logging. For example, a transportation company may monitor the latency and throughput of its real-time GPS tracking pipeline and configure alerts for any data loss or delays.

Armed with a structured approach to designing data pipelines, you'll be better prepared to tackle technical interview questions. Let's proceed to dissect some questions to help you fully prepare for your interviews.

Technical interview questions

In this section, we will prepare you for technical interview questions specifically focused on data pipeline design. These questions aim to assess your understanding of the concepts and practical considerations involved in designing efficient and reliable data pipelines:

*   *Question 1*: What is the difference between ETL and ELT?

    Answer: ETL involves the extraction of data from source systems, its transformation into a usable format, and its loading into a target database or data warehouse. In contrast, ELT involves extracting data and loading it into the target system before transformation. ELT is typically more effective when the target system is robust enough to handle transformations quickly, such as modern cloud-based data warehouses such as Snowflake or BigQuery.

*   *Question 2*: How would you ensure data quality in your pipeline?

    Answer: Data quality can be maintained by incorporating validation checks at various pipeline stages. For instance, sanity checks and business rule validations can be applied immediately after data extraction and transformation, respectively. In addition, automated testing and monitoring can be configured to flag any data anomalies or inconsistencies.

*   *Question 3*: How would you handle data loss in a stream processing pipeline?

    Answer: Using a dependable message broker that supports data replication, such as Kafka, can minimize data loss. Checkpointing mechanisms can be implemented to periodically save the stream's state. In the event of data loss, the system can revert to its last known good state and reprocess the lost information.

- *Question 4*: How do you deal with schema changes in source systems?

  Answer: Incorporating flexibility in the pipeline to adapt to the new schema, utilizing a schema registry, or versioning the data can be deployed to manage schema changes. Notifications or alerts can be configured to inform of such changes, enabling proactive pipeline adjustment.

- *Question 5*: How do you ensure **fault tolerance** (**FT**) in a data pipeline?

  Answer: Several techniques, including data replication, checkpointing, and automatic retries, can be utilized to achieve FT. Using distributed systems with inherent FT, such as Hadoop and Spark, can also be beneficial. Monitoring and alerting can be set up to detect failures quickly and initiate recovery procedures.

Summary

In this chapter, we explored the intricacies of data pipeline design for data engineers. We covered the foundational concepts of data pipelines and the step-by-step process of designing pipelines and prepared you for technical interview questions related to data pipeline design.

By understanding the fundamentals, following best practices, and showcasing your expertise in data pipeline design, you will be well prepared to architect, implement, and maintain efficient and reliable data pipelines. These pipelines serve as the backbone for data processing and analysis, enabling organizations to leverage the power of their data.

In the next chapter, we will delve into the exciting field of data orchestration and workflow management. We will explore tools, techniques, and best practices for orchestrating complex data workflows and automating data engineering processes. Get ready to streamline your data operations and enhance productivity as we continue our journey into the world of data engineering!

12

Data Warehouses and Data Lakes

In today's data-driven society, organizations are constantly seeking methods to manage, process, and extract insights from vast quantities of data. Traditional databases and processing systems often fall short in terms of scalability, flexibility, and performance as data volume, velocity, and variety increase. Various systems and architectures are utilized by businesses for different purposes, ranging from transactional data in databases to big data in cloud storage. You'll probably come across two main categories of data storage architectures: data lakes and warehouses.

We will delve into these two fundamental components in this chapter. You will discover the underlying ideas that distinguish them, the subtleties of their architecture, and the data flows that occur both inside and between these systems.

In this chapter, we will cover the following topics:

- Exploring data warehouse essentials for data engineers
- Examining data lake essentials for data engineers
- Technical interview questions

Exploring data warehouse essentials for data engineers

Data warehouses are the backbone of modern data analytics. They are a combination of intricate architecture, careful data modeling, and effective processes that ensure data is not only stored but is also accessible, consistent, and meaningful. Let's look at these essentials in the next subsections.

Architecture

It's similar to knowing the blueprints of a complicated building to comprehend the architecture of a data warehouse. We'll break down the different layers that comprise a typical data warehouse architecture in this section.

The source layer is where data originates in a data warehouse and is in its original, unmodified state. A variety of data types, including flat files, external APIs, and databases, can be included in this layer. Making sure this layer is both easily and securely accessible for data ingestion is the first task facing you as a data engineer. After being extracted, the data moves to the staging area, a short-term storage space meant to hold it while it goes through the crucial **Extract**, **Transform**, **Load** (**ETL**) procedure. In this case, you may encounter issues such as inconsistent data originating from different sources.

The center of the data warehouse architecture is undoubtedly the ETL layer. Here, data is standardized, cleaned, formatted, enriched, and aggregated. For instance, you may have to standardize comparable measurements, such as pounds and kilograms, into a single unit or impute missing values. The data is moved into the data storage layer after transformation. This layer is where you'll use data modeling techniques because it's optimized for efficient querying. Data is frequently arranged using well-known schemas such as star or snowflake schemas to enhance query performance.

Lastly, end users can interact with the warehouse data through the data presentation layer. This layer could have summarized reports, compiled tables, and dashboards. During this last stage, as a data engineer, you frequently work with data analysts to make sure that the data presented is clear and easy to understand in addition to meeting business requirements.

Here are the different stages of the ETL process in more detail:

- **The ETL process**: The ETL process is a key part of data warehousing, and it's crucial to understand this when working with data at scale. Here's an explanation of each step:

  - **Extract**: The first step in the ETL process involves extracting data from various sources. The data may come from relational databases, flat files, web APIs, and numerous other places. The main challenge in this step is dealing with the different formats, schemas, and quality of data. This requires developing connections to these various systems, understanding their data models, and extracting the needed data in a way that minimally impacts the performance of the source systems.

  - **Transform**: In the transformation step, the raw data extracted from the source systems is cleaned and transformed into a format that's suitable for analytical purposes. This often involves a variety of transformations, such as the following:

    - **Cleaning**: Cleaning is the process of dealing with `null` or missing values, getting rid of duplicates, or fixing values that are out of range

    - **Standardizing**: Data conversion to a common measurement system, text value standardization (for example, mapping `true`/`false` and `yes`/`no` to a common Boolean representation), and date and time format standardization are examples of standardizing

    - **Enrichment**: This could involve adding extra data, such as calculating new measures or adding geo-coordinates for addresses

- **Reformatting**: This could include reorganizing the data into different formats, such as splitting or merging columns or altering the data types

- **Aggregation**: This phase could involve computations such as averaging total sales by area

The aim is to ensure the quality and consistency of data, making it easier to work with in the data warehouse.

- **Load**: The load step involves writing the transformed data to the target database, which is often a data warehouse. This step needs to be carefully managed to ensure data integrity and minimize the impact on the system, which needs to remain available to users during the loading process.

There are generally two types of loading strategies:

- **Full load**: As the name suggests, the entire dataset is loaded in one batch. This is typically done when a new ETL process is established.

- **Incremental load**: Here, only data that has changed since the last load is updated. This is a more common and efficient approach, especially when dealing with large volumes of data.

In some systems, the loading process also involves building indexes and creating partitioning schemes to improve query performance.

The ETL process is at the heart of data engineering, and understanding it is critical for managing and maintaining robust data pipelines. It's through this process that raw data is converted into meaningful information that can be used for business analysis and decision-making.

- **Data modeling**: The process of creating a visual representation of how data will be stored in a database is known as data modeling. It is a conceptual representation of the associations between various data objects, the rules governing these associations, and the data objects themselves. Data modeling facilitates the visual representation of data and ensures that it complies with legal requirements, business rules, and governmental directives.

Data elements are defined and organized using data models, along with their relationships to one another and to the characteristics of real-world entities. This can be done for a number of purposes, such as database data organization, software structure creation, or even network architecture creation. There are several types of data models:

- **Conceptual data model**: In data modeling, this is the highest level of abstraction. Without going into technical details, it provides a broad overview of interactions between various entities in the business domain. Before delving into the technical specifics, a data engineer may develop a conceptual model to convey to stakeholders the general structure of the data and make sure everyone is in agreement.

- **Logical data model**: Through the addition of structure and limitations, such as data types and table relationships, this model expands upon the conceptual model. It is not dependent on any one technology, even though it has more detail than the conceptual model. The maintenance of data integrity and relationships during a data warehouse migration from one platform to another, such as from on-premises to the cloud, depends heavily on the logical data model, which acts as an intermediary step.

- **Physical data model**: All of the implementation details unique to each database are included in the physical data model. It outlines the methods for storing, indexing, and retrieving data while optimizing performance. In order to optimize query performance, you may need to decide whether to use columnar storage or how to partition tables.

When establishing or optimizing a data warehouse, having a thorough understanding of the subtle differences between these various models can help you make better decisions. Your decision will affect how well data can be queried, how simple it is to understand, and how easily it can be modified to meet changing requirements, regardless of whether you are building new models from scratch or altering ones that already exist.

As we proceed, let's concentrate on the nuances of schemas, which are an additional essential component that support data modeling in the creation of an effective data warehouse.

Schemas

Within a data warehouse, schemas serve as the structural design frameworks that impact how data is arranged, saved, and accessed. They are just as important in determining a data warehouse's usefulness and effectiveness as architectural plans are in determining a building's appearance and functionality. Now, let's explore the primary schema types that are frequently used in data warehousing.

Two main types of schemas that you will encounter in data engineering are star and snowflake schemas. Let's look at these in more detail:

- **Star schema**: With a central fact table that is directly connected to multiple dimension tables, this is the most basic type of dimensional model. Because a star schema offers a simple method of data organization and permits the joining of fewer tables, it is typically simple to comprehend and effective for querying. For instance, a star schema's simplicity and performance can often come in handy if you're tasked with developing a quick reporting solution for a small-to-medium-sized business.

The following diagram displays the general architecture of a star schema:

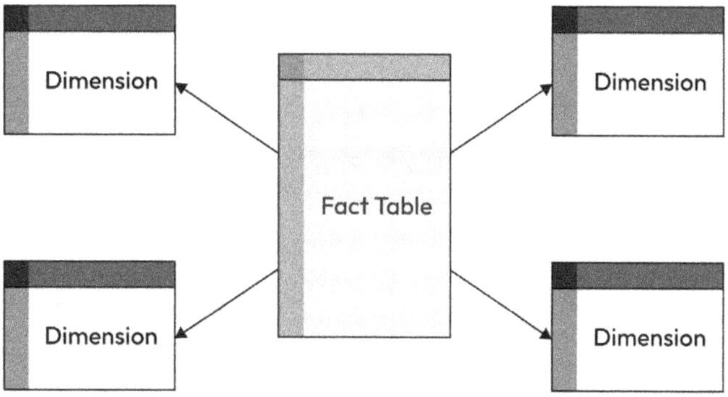

Figure 12.1 – Example of a star schema

Now let's understand the architecture seen in the preceding diagram in the following bullets:

- **Fact tables**: Fact tables hold the data to be analyzed, and they summarize an enterprise's business activities or transactions. These are quantifiable data or metrics that can be analyzed. For example, in a retail business, a fact table might include fields such as items sold, the total amount of sales, the number of items sold, and so on.

- **Dimension tables**: Dimension tables contain the textual context of measurements captured in fact tables. They contain details used to query, filter, or classify facts. For example, a dimension table for customers may contain fields such as Customer Name, Address, Email, and so on.

- **Snowflake schema**: A snowflake design divides dimension tables into more normalized tables by normalizing them, in contrast to a star schema. Although this reduces data redundancy, query performance suffers as a result of the additional joins required to obtain the same data. If you are working with a complex database where normalization and data integrity are more important than query speed, you may choose to use a snowflake schema.

The following diagram provides an example of a snowflake schema:

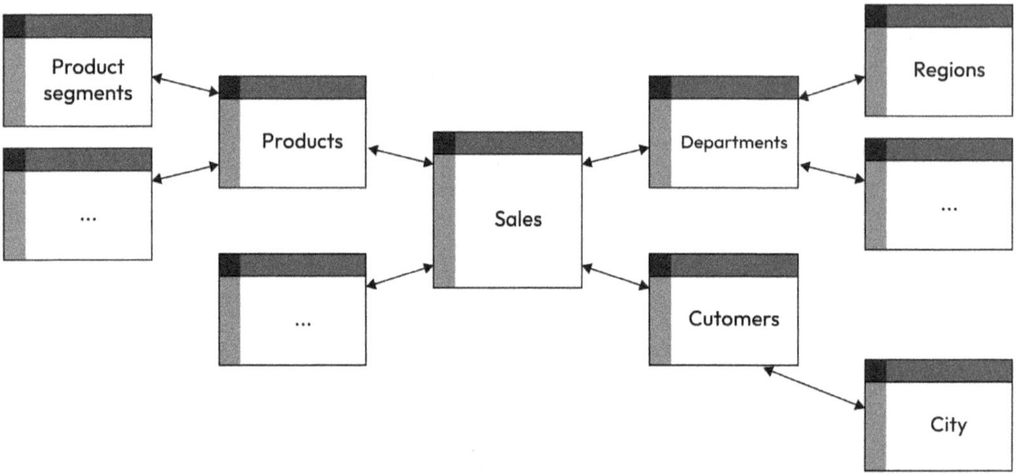

Figure 12.2 – Example of a snowflake schema

Schemas have an impact on data modeling, ETL procedures, and the storage layer itself. Finding the right schema frequently requires striking a balance between query performance optimization and data redundancy reduction. Your decision will be influenced by a number of variables, including the particular business requirements, the type of data, and the anticipated workload from queries.

Understanding the distinctions and ramifications of these schema types gives you access to an additional level of knowledge that is crucial for data warehousing. With this knowledge, you can make well-informed decisions that will ultimately affect how well data is stored and retrieved, which will affect how well your organization's **business intelligence (BI)** initiatives succeed.

Examining data lake essentials for data engineers

Large volumes of unprocessed data can now be cost-effectively and flexibly stored with data lakes. We'll examine the layers and architecture of data lakes in this section, explaining their differences from data warehouses and situations in which using one is preferable. We'll explore the key areas of a data lake, discussing their special features and the kinds of situations in which a data engineer could interact with them.

Data lake architecture

A data lake's architecture encompasses more than just the zones that make up the lake. It is a combination of different parts that make data intake, storage, processing, and consumption easier. Gaining an understanding of the architecture will enable you to create and run a data lake efficiently as a data engineer. Let's examine the essential components of a data lake architecture in more detail:

- **Data lake zones**: Data lake zones are a way of organizing data within a data lake. Each zone serves a specific purpose and follows specific rules about which data it can contain and how that data can be used. Here's a detailed description of the different zones often found in a typical data lake:

  - **Raw zone (or landing zone)**: This is the initial area where data lands when it first arrives in the data lake from various data sources. The data in this zone is in its raw, unprocessed form and maintains its original structure. The raw zone serves as an immutable, historical record of all data ingested into the data lake. This is particularly useful for audit purposes and troubleshooting data ingestion issues.

  - **Clean zone**: In the clean zone, data is cleansed, validated, and transformed to a common format, making it more consumable. Typical actions in this zone include data type conversions, `null` value handling, validation against business rules, deduplication, and standardization of values. This zone ensures that the data is clean and prepared for further transformations or analysis.

  - **Curated zone (or business zone)**: This zone contains data that has been further transformed and enriched, often to conform to the organization's standard data model or schema. The transformations in this zone are usually more business-specific and can involve combining data from multiple sources, aggregating or disaggregating data, applying business rules, and so on. Data in the curated zone is typically used for business reporting and **business analytics (BA)**.

  - **Sandbox zone (or exploration zone)**: This zone is intended for data exploration and data science activities. Here, data scientists and analysts can experiment with the data, build and test new models, or explore new business questions. This zone provides a space where users can work with the data without impacting other zones.

  - **Secure/restricted zone**: This zone contains sensitive or regulated data that is subject to stricter access controls and governance rules. It's often encrypted, and access is granted on a need-to-know basis. This zone ensures that sensitive data is protected and used responsibly.

These zones provide a structured way to manage data within a data lake. They ensure that there are areas to store raw data, clean and process it, perform more business-specific transformations, and even experiment with new ideas, all while maintaining the security of sensitive data. By understanding and implementing these zones, data engineers can create a more organized and manageable data lake.

- **Data lake processing**: Data lake processing is a crucial step that prepares the raw, ingested data into a form that can be used for data analysis, reporting, and **machine learning** (**ML**). Data lake processing usually involves a few key steps:

  - **Data cleansing**: The process of finding and fixing errors in data.

  - **Data transformation**: The process of changing data from one format or structure to another is known as data transformation. This might mean combining data, changing the types of data, standardizing or normalizing values, and so on. Data transformation prepares the data for specific uses, such as analysis or ML.

  - **Data enrichment**: Enhancing, honing, or otherwise improving raw data is the goal of data enrichment. This could entail combining data from several sources, extracting new data attributes, or adding more information from outside sources. Data enrichment makes data more analytically useful by providing additional context.

  - **Data cataloging**: Making a searchable inventory of data assets in the data lake is the process of data cataloging. Documenting the metadata for data assets, such as the type, source, quality, and any transformations made to the data, is part of this. Users can locate and comprehend data in the data lake with the aid of data cataloging.

Data governance and security

A vital component of data management is security and governance, especially for large-scale data systems such as data lakes and warehouses. They are designed to guarantee that information is kept up to date, used responsibly, and shielded from intrusions and breaches.

Data governance refers to the overall management of the availability, usability, integrity, and security of data employed in an enterprise. It's a collection of practices and guidelines aimed at ensuring high data quality, consistency, and reliability in the organization. The main components of data governance include the following:

- **Data quality**: Ensuring the completeness, accuracy, and consistency of data throughout its entire life cycle.

- **Data lineage:** The process of tracking data from its source to its destination and analyzing changes it undergoes over time. This is especially crucial for regulatory compliance, auditing, and troubleshooting.

- **Metadata management**: The process of managing your data's metadata, which can include details about the origin, transformations, quality, and other aspects of your data.

- **Data access and privacy policies**: These include providing information about who is authorized to access which data and how, in compliance with privacy laws and regulations, they can use it.

Data security

As part of data security, protective digital privacy measures are put in place to prevent unauthorized access to computers, websites, and databases. This is necessary for maintaining data integrity, avoiding data breaches, and abiding by privacy laws and regulations. Some crucial components of data security in a data lake are listed as follows:

- **Access control**: This involves managing who has access to which data. This could include implementing **role-based access control (RBAC)** or **attribute-based access control (ABAC)**.

- **Encryption**: This involves encoding data to protect it from unauthorized access. Data can be encrypted at rest (when it's stored) and in transit (when it's being moved from one place to another).

- **Data masking and anonymization**: This involves concealing private or sensitive information within your dataset to protect individuals' privacy.

- **Security monitoring and auditing**: This involves tracking and monitoring data usage and access patterns to detect any potential security threats or breaches.

Data engineers need to understand data governance and security to design and manage data systems that are reliable, consistent, and secure. These practices help ensure that data is a trusted and protected asset in the organization.

Mastering these essential concepts will prepare you to discuss data lake principles in technical interviews. You will gain the necessary knowledge to design scalable data storage architectures, implement data processing frameworks, and ensure proper data governance and security within the data lake environment.

Technical interview questions

Now that we have explored essential concepts of data warehouses and data lakes, it's time to put your knowledge to the test with technical interview questions. This section will cover a range of interview questions that assess your understanding of data warehouse and data lake principles. As we dive into these questions, prepare to showcase your expertise and problem-solving skills:

- *Question 1*: What is the primary purpose of a data warehouse?

  Answer: The primary purpose of a data warehouse is to store, organize, and consolidate large volumes of structured and historical data to support BI and data analytics.

- *Question 2*: How does a data lake differ from a data warehouse?

  Answer: Unlike a data warehouse, a data lake stores data in its raw format without the need for predefined schemas. It accommodates various data types and allows more flexible data exploration and analysis.

- *Question 3*: What is the role of ETL processes in a data warehouse?

  Answer: ETL processes extract data from various sources, transform it into a consistent format, and load it into the data warehouse for further analysis and reporting.

- *Question 4*: How would you handle data integration challenges when populating a data warehouse?

  Answer: Data integration challenges can be addressed through data mapping, data cleansing, data transformation, and ensuring data quality through validation and error handling.

- *Question 5*: What are the benefits of dimensional modeling in data warehousing?

  Answer: Dimensional modeling simplifies complex data structures, improves query performance, and enables efficient data analysis by organizing data into fact and dimension tables.

- *Question 6*: How can you optimize query performance in a data warehouse?

  Answer: Query performance can be optimized by creating appropriate indexes, implementing partitioning strategies, optimizing data loading processes, and tuning the database configuration.

- *Question 7*: What are the advantages of using a distributed filesystem for data lake storage?

  Answer: Distributed filesystems offer scalability, **fault tolerance** (**FT**), and the ability to handle large volumes of data across a cluster of machines, making them ideal for storing and processing data in data lake environments.

- *Question 8*: How can data governance be ensured in a data lake?

  Answer: Data governance in a data lake can be confirmed by implementing access controls, data classification, metadata management, data lineage tracking, and adhering to regulatory compliance requirements.

- *Question 9*: What are the differences between batch processing and real-time streaming in data lakes?

  Answer: Batch processing involves processing data in large volumes at scheduled intervals, while real-time streaming processes data in near real time as it arrives. Batch processing is suitable for analyzing historical data, while real-time streaming enables immediate insights and responses.

- *Question 10*: How can you handle schema evolution in a data lake?

  Answer: Schema evolution in a data lake can be managed through schema-on-read, which is applied during data access, enabling flexibility and accommodating changes in data structure over time.

- *Question 11*: What is the role of metadata in a data lake?

  Answer: Metadata provides essential information about data stored in a data lake, including data lineage, data quality metrics, data schemas, and data usage patterns. It facilitates data discovery and enhances data governance and data management processes.

- *Question 12*: How would you ensure data security in a data lake environment?

  Answer: Data security in a data lake can be confirmed through access controls, encryption techniques, user authentication and authorization mechanisms, and adherence to privacy regulations and data protection policies.

- *Question 13*: What are the benefits of using columnar storage in a data lake?

  Answer: Columnar storage in a data lake offers several benefits, including improved query performance due to column-level compression and encoding, efficient data compression, and the ability to access specific columns without reading unnecessary data. It also enables faster data processing and analysis for analytics workloads.

- *Question 14*: How would you handle data quality issues in a data lake?

  Answer: Data quality issues in a data lake can be addressed through data profiling, validation, and cleansing techniques. Implementing data quality checks, monitoring data quality metrics, and establishing data governance practices are essential for maintaining data quality.

- *Question 15*: Can you explain the concept of data lineage in a data lake?

  Answer: Data lineage in a data lake refers to the ability to track the origin, transformations, and movement of data throughout its life cycle within the data lake. It helps understand data dependencies, ensure data accuracy, and comply with data governance requirements.

By familiarizing yourself with these technical interview questions and their answers, you will be well prepared to showcase your expertise in data warehousing and data lakes. Remember to focus on the correct answers and demonstrate your problem-solving skills, critical thinking, and ability to apply these concepts to real-world scenarios.

The next chapter will explore the exciting realm of **continuous integration/continuous development (CI/CD)** for data engineering. Get ready to explore principles and practices that enable efficient and reliable software development in the data engineering context. Let's continue our journey toward mastering the essential skills of a data engineer.

Summary

This chapter covered data warehouses and data lakes, essential tools for data engineers. We studied these systems' architecture, operation, and best practices. We started with data warehouses and how they use data marts and schemas to analyze structured transactional data. We examined their layered architecture and the ETL process, which underpins data warehouse operations.

Data lake architecture—from data ingestion and storage to data processing and cataloging—was our next topic. We explained data lake zones and their importance to organization and functionality. The difference between a well-managed data lake and a data swamp and the importance of data governance and security was stressed.

The next chapter will explore data engineering's exciting CI/CD world. Prepare to learn about data engineering software development principles and practices that ensure efficiency and reliability. Let's keep learning data engineering skills.

Part 4:
Essentials for
Data Engineers Part III

In this part, we will provide an overview of additional topics with extra interview questions to further prepare you for your interview.

This part has the following chapters:

13

Essential Tools You Should Know

As data engineers, we rely on a myriad of software tools to process, store, and manage data effectively. In this chapter, we will explore the essential tools every data engineer should know. These tools will empower you to harness the power of the cloud, handle data ingestion and processing, perform distributed computations, and schedule tasks with efficiency and precision. By the end of this chapter, you'll have a strong understanding of the key tools in data engineering, along with the knowledge of where and how to apply them effectively in your data pipeline.

In this chapter, we will cover the following topics:

- Understanding cloud technologies
- Mastering scheduling tools

Understanding cloud technologies

Cloud technologies provide the fundamental framework for a wide range of data engineering tasks in today's data-driven world. Cloud platforms provide the scalability, reliability, and flexibility that modern enterprises require, from data collection to processing and analytics. This section gives a brief introduction to cloud computing, outlines the main products and services offered by top cloud providers such as AWS, Azure, and Google Cloud, and goes into detail about key cloud services that are critical to data engineering. Additionally, you'll discover how to assess cloud solutions according to the most important factors for your data engineering projects, including cost-effectiveness, scalability, and dependability. Understanding the fundamentals of cloud computing will help you make informed decisions in real-world data engineering scenarios and prepare you for the inevitable cloud-related interview questions.

Get ready to assemble your toolbox, tailored for success in your data engineering endeavors.

Major cloud providers

Three companies routinely rule the vast ecosystem of cloud technologies: **Google Cloud Platform (GCP)**, Microsoft Azure, and **Amazon Web Services (AWS)**. All of these service providers provide an extensive range of services designed to meet various data engineering requirements:

- *Amazon Web Services*: With a wide range of services, ranging from basic computing and storage options such as EC2 and S3 to more specialized offerings for big data, analytics, and machine learning, AWS has long been a pioneer in the cloud space. Because of its extensive global infrastructure and mature platform, AWS is the preferred option for organizations seeking scalability and reliability.

- *Microsoft Azure*: Another major player in the cloud space, Microsoft Azure makes use of its integrations with Windows and other Microsoft enterprise software. Azure offers many different types of cloud services, such as networking, analytics, computing, and storage. It is a formidable competitor due to its ability to integrate with current Microsoft-based enterprise environments, particularly for businesses that have a strong foundation in the Microsoft ecosystem.

- *Google Cloud Platform*: GCP excels in open source technologies, machine learning, and data analytics. GCP is well known for its high-performance computing, data storage, and data analytics services. It is built on the same infrastructure that enables Google to deliver billions of search results in milliseconds, serve six billion hours of YouTube video each month, and provide storage for one billion Google Drive users.

The choice you make will often rely on the particular requirements, financial constraints, and strategic objectives of your project or organization. Each of these major cloud providers has advantages and disadvantages of its own.

After examining the main cloud providers, let's take a closer look at the essential services that these companies provide for data engineering projects.

Core cloud services for data engineering

Whichever major cloud provider you select, three essential service categories—compute resources, storage solutions, and networking—become especially important when it comes to data engineering. Having a firm grasp of these core services will help you when designing cloud-based data solutions:

- *Compute resources*: The foundation of any data engineering pipeline is compute resources. AWS's EC2, Azure's virtual machines, and GCP's Compute Engine are just a few of the compute services that cloud providers offer. These services can be scaled up or down based on your processing requirements. Your data ingestion engines, transformation jobs, and analytics algorithms depend on these resources to function. They are available in a variety of configurations, suited for different types of computing requirements, ranging from memory-optimized processes to CPU-intensive tasks.

- *Storage solutions*: These are yet another crucial element. Having dependable and scalable storage is essential, regardless of whether you're storing finalized analytics reports, intermediate datasets, or raw data. Highly reliable and readily available storage solutions that can be adjusted for various accessibility and retention needs can be found with services such as AWS S3, Azure Blob Storage, or Google Cloud Storage. Furthermore, certain providers provide analytics workload-optimized big data storage solutions, such as AWS Redshift or Google BigQuery.

- *Networking*: Cloud platforms with networking capabilities make it easier to move data quickly and securely between different parts of your architecture. Customizable network configurations are made possible by services such as AWS VPC, Azure Virtual Network, and Google VPC. These configurations enable secure, isolated data flows as well as smooth integration with on-premises or other cloud-based resources. Having efficient networking guarantees that your data is not only safe to process but also easily accessible, which results in an optimized data pipeline.

Having a comprehensive understanding of these fundamental cloud services will put you in a better position to create reliable data engineering solutions. Let's now turn our attention to the tools that are essential for efficient data processing, storage, and ingestion, as well as how to choose the best ones for your projects.

Identifying ingestion, processing, and storage tools

Choosing the appropriate tools for data ingestion, processing, and storage is a logical next step in developing a strong data engineering pipeline, after laying the foundation with cloud technologies. These instruments are the gears in the machine that guarantee the smooth movement, handling, and preservation of data. The purpose of this section is to familiarize you with common tools for each of these essential tasks. Gaining an understanding of the features, advantages, and disadvantages of these tools will enable you to design scalable and effective data pipelines, making you an excellent prospect in any data engineering interview and a priceless asset in practical situations:

- *Apache Kafka*: For real-time data ingestion, Kafka has established itself as the industry standard, especially in situations requiring the management of enormous data streams. Designed by LinkedIn and subsequently made available to the public, Kafka provides fault tolerance, high throughput, and real-time analytics and monitoring. Because of its publish-subscribe architecture, data pipelines can be separated, offering more scalability and flexibility. Kafka is very extensible and can be integrated with a wide range of data sources and sinks thanks to its rich ecosystem, which consists of numerous connectors and APIs.

- *Apache Flume*: Another popular tool for data ingestion is Flume, which is specially designed for ingesting data into Hadoop environments. Large volumes of log data or streams of event data from multiple sources can be gathered, aggregated, and moved using Flume to a centralized data store such as HDFS. It has a strong, fault-tolerant architecture that can manage large amounts of data. Even though Flume isn't as flexible as Kafka, it's easy to set up and works especially well for certain use cases, such as aggregating log or event data.

Both Kafka and Flume have advantages of their own, and which one you choose will mostly depend on your needs when it comes to data intake, such as the kinds of data sources you work with, the amount of data, or the requirement for real-time processing.

Now that we've discussed the fundamental data ingestion tools, let's examine the technologies needed to efficiently process the ingested data.

Data storage tools

Data must be stored effectively and dependably in order to remain safe and accessible after it has been ingested and processed. Depending on the type, volume, and future uses of the data, storage requirements can vary greatly. Next, we'll go over a few of the common storage options that you'll probably run into in the data engineering field:

- **HSFS**: Many big data projects are built around the open source HDFS storage system. Built to function in concert with the Hadoop ecosystem, HDFS offers scalable, fault-tolerant storage that is capable of handling petabytes of data. It allows you to distribute your data across several machines and performs well in a distributed environment. When combined with other Hadoop ecosystem tools such as MapReduce for data processing, HDFS is especially useful for projects that need high-throughput access to datasets.

- **Amazon S3**: Amazon Simple Storage Service, is a flexible cloud storage option with excellent scalability, availability, and durability. Large multimedia files and small configuration files can both be stored on it, and it can also be used as a data lake for big data analytics. S3 is frequently used to store important business data because of its strong security features, which include data encryption and access control methods.

- **Google Cloud Storage**: Like Amazon S3, Google Cloud Storage provides an extremely scalable, fully managed, and adaptable object storage solution. It works well for storing data for disaster recovery and archiving purposes, delivering big data objects to users over HTTP or HTTPS, and serving both structured and unstructured data.

- **Microsoft Azure Data Storage**: Azure, from Microsoft, provides a variety of storage services to meet various storage needs, such as Blob Storage, Azure Files, and Azure Queues. While Azure Files is designed for file sharing and Azure Queues is useful for storing and retrieving messages, Azure Blob Storage is especially helpful for storing unstructured data. Azure is a desirable option for businesses that have already invested in Microsoft technologies because of its ability to integrate with Microsoft's software stack.

The choice between these storage options will rely on a number of factors, including data type, access patterns, and enterprise-specific requirements. Each of these options has a unique set of features.

Now that we have a solid understanding of data storage options in place, let's turn our focus to scheduling tools, which aid in the coordination and automation of the whole data pipeline.

Mastering scheduling tools

Coordinating these elements into a smooth, automated workflow is crucial after you've set up your data engineering environment with the right ingestion, processing, and storage tools. Scheduling tools are useful in this situation. These tools control how jobs and workflows are carried out, making sure that things get done in the right order, at the right time, and in the right circumstances. This section will walk you through the features, use cases, and comparative analysis of some of the most widely used scheduling tools, including Luigi, Cron Jobs, and Apache Airflow. Equipped with this understanding, you will be capable of efficiently designing and overseeing intricate data pipelines—a capability that is not only essential for job interviews but also highly valuable in practical settings.

Importance of workflow orchestration

Beyond just carrying out tasks at predetermined times, scheduling serves other purposes as well. It entails intricate workflow orchestration in which jobs are interdependent and must be successfully completed before starting another. For example, before data transformation can start, a data intake task needs to be finished successfully. Data transformation must then be finished before data analysis can start. The pipeline's seamless operation and the preservation of data integrity are guaranteed by this sequencing. Additionally, failure recovery, logging, monitoring, and notifications are features that modern schedulers provide and are essential for preserving stable and dependable data pipelines.

Apache Airflow

Across a wide range of industries, Apache Airflow has become the de facto standard for orchestrating complex data pipelines. A highly valued feature in data engineering, Airflow was first created by Airbnb and subsequently contributed to the open source community. It offers a high degree of customization and flexibility.

Now that we've explored the nuances of scheduling tools, let's review what we've discovered and talk about how you can use this knowledge to succeed in your data engineering career and ace interviews.

At its core, Airflow's architecture consists of several components:

- *The scheduler*: In charge of arranging the execution of tasks according to their dependencies
- *Worker nodes*: These carry out your job's tasks and notify the scheduler of its progress
- *Metastore database*: Tasks, their statuses, and other metadata are tracked in the metastore database
- *Web server*: Offers a graphical user interface for tracking workflow, resuming abandoned tasks, and initiating spontaneous runs

Key features of Airflow include the following:

- **Directed Acyclic Graphs (DAGs)**: Airflow workflows are defined as code that takes the shape of DAGs, which enable pipeline construction that is dynamic

- **Extensibility**: It is very flexible due to the large number of plugins and integrations that are available, including hooks to well-known data sources and sinks

- **Logging and monitoring**: Airflow has strong logging features, and its web-based interface offers choices for monitoring jobs in real time

Since Apache Airflow is widely used in industry and is often asked about in interviews, learning about it is essential to master modern data engineering practices.

Being flexible and extensible, Apache Airflow is one of the most widely used scheduling tools in modern data engineering. Airflow, developed by Airbnb and later made available to the public, lets you specify intricate processes using code, usually in the form of Python. Numerous features are available, including logging, error handling, dynamic pipeline creation, and an advanced monitoring user interface. Because of its extensive plugin and integration ecosystem and community-driven development, Airflow is highly customizable to meet a wide range of data engineering requirements.

One of the earliest and most basic scheduling tools is cron, which is especially well liked on Unix-based systems. Despite not having all the bells and whistles of more contemporary solutions, its lightweight design and simplicity make it appropriate for simple tasks. Cron Jobs are typically used to schedule repetitive tasks such as basic reporting functions, data backups, and simple data transformations. It is important to remember, though, that Cron is not intended to manage task dependencies or offer features such as logging and monitoring right out of the box.

The choice will mostly depend on the complexity of your data pipelines, your familiarity with programming languages, and particular workflow requirements.

Summary

Well done on learning about the key resources that each and every data engineer should be aware of! Cloud applications, data ingestion, processing, storage tools, distributed computation frameworks, and task scheduling solutions were all covered in this chapter. You've given yourself a strong toolkit to tackle a variety of data engineering challenges by becoming acquainted with these tools.

Recall that becoming an expert with these tools is only the start of your career as a data engineer. Your ability to adjust to the constantly changing data landscape will depend on your continued exploration of and adherence to emerging technologies and tools. Take advantage of these tools' opportunities to advance your data engineering abilities.

In the next chapter, we will explore the world of **continuous integration/continuous development (CI/CD)**.

Continuous Integration/ Continuous Development (CI/CD) for Data Engineers

It takes more than just mastering a set of techniques to succeed in the field of data engineering. You must keep up with the rapidly changing environment's new tools, technologies, and methodologies. The fundamental principles of **continuous integration** and **continuous development (CI/CD)**, which are crucial for any data engineer, are the focus of this chapter.

Understanding CI/CD processes will give you a versatile skill set that will not only increase your effectiveness in your current position but also have a big impact on the performance and dependability of the systems you create. In this chapter, you'll learn how to use Git for version control, gain insight into fundamental automation concepts, and develop your skills in building robust deployment pipelines. You'll comprehend by the end of this chapter why these abilities are essential for upholding a high level of quality, dependability, and effectiveness in the constantly developing field of data engineering.

In this chapter, we're going to cover the following topics:

- Essential automatic concepts
- Git and version control
- Data quality monitoring
- Implementing continuous deployment

Understanding essential automation concepts

One of the pillars of effective, dependable, and scalable data engineering practices is automation. Manual interventions not only increase the chance of error in today's quick development cycles, but are also becoming increasingly impractical given the size and complexity of today's data systems.

The purpose of this section is to acquaint you with the fundamental automation ideas that form the cornerstone of a well-executed CI/CD pipeline.

We'll examine the three main types of automation—test automation, deployment automation, and monitoring—to give you a comprehensive understanding of how these components interact to speed up processes and guarantee system dependability. To create systems that are not only functional but also reliable and simple to maintain, you must master these automation techniques, whether you're creating a real-time analytics engine or setting up data pipelines for machine learning.

Test automation

Tests are written before the necessary code in the **test-driven development** (TDD) method of software development. Before implementing the actual code, data engineers may need to write tests to verify data schemas, transformation logic, or even data quality. This procedure is essential for finding errors at their root, facilitating the development of a stable code base, and minimizing the amount of debugging required in the future. Consider being in charge of a data pipeline that integrates data from various sources into a single data warehouse. By using TDD, you can make sure that data is properly formatted, transformed, and loaded while also spotting any errors early on in the development cycle.

Different test types have different purposes in the field of data engineering:

- **Unit tests**: These tests concentrate on particular functions for data transformation in your application. To check whether a function accurately converts temperature values from Fahrenheit to Celsius, for instance, you might create a unit test.

- **Integration tests**: These tests verify how different parts of your system interact with one another. Think about how you might have constructed two distinct data ingestion pipelines, one for customer data and one for transaction data. These pipelines could be properly fed into a combined customer-transaction database through an integration test.

- **End-to-end tests**: These tests examine how a whole procedure or workflow flows, typically by simulating real-world situations. To ensure that your real-time analytics engine is successfully ingesting data streams, carrying out the necessary transformations, and updating dashboards without any glitches, for example, you could run an end-to-end test.

In a data engineering context, a variety of frameworks and tools can make automated testing easier. Writing unit and integration tests is made very flexible and simple by Python libraries such as `pytest`. Tools such as Great Expectations and **Data Build Tool** (**dbt**) can be used for data pipeline testing to verify whether your data is properly ingested and transformed. In order to validate the entire data processing workflow, end-to-end testing solutions such as Selenium can automate user-like interactions with your analytics dashboard.

By embracing test automation, you give yourself a potent tool for ensuring the accuracy of the code and the dependability of the system. Knowing you're not adding new flaws to the current system gives you the assurance to make changes and improvements. Automated testing will be your steadfast ally in upholding a high standard of quality, whether you're modifying a challenging **Extract**, **Transform**, **Load** (**ETL**) job or scaling your data pipelines.

You now have a thorough understanding of test automation and its crucial function in data engineering. Let's move on to the following crucial automation component: deployment automation. In order to make sure that your data pipelines and systems are consistently dependable and easily scalable, this will go into detail about how to automate the process of moving your code from development environments to production environments.

Deployment automation

For many reasons, automating the deployment process is crucial. The risk of human error such as incorrect configurations or skipped steps, which can cause instability in your systems, is reduced first and foremost. Second, automated deployments typically occur more quickly and with greater reliability than manual ones, enabling shorter release cycles. Finally, automation makes sure that your process is repeatable and scalable, which is important for systems that frequently need to grow or change. This is especially true in the field of data engineering, where pipelines and data processes must be both dependable and adaptable in order to meet changing business needs and data source requirements.

There are numerous tools available to help with deployment automation, each with unique advantages and applications. Jenkins is a well-liked option due to its flexibility and robust ecosystem of plugins. Other choices, such as GitLab CI/CD and CircleCI, are renowned for their simple version control system integration. Kubernetes provides robust container orchestration for those interested in cloud-native solutions, and tools such as Spinnaker can help manage intricate deployment pipelines.

Several tactics can be used to control how your application is made available to end users:

- **Blue-green deployment**: This strategy uses a blue environment for the current version and a green environment for the new version. Traffic is switched from blue to green after the green environment has been thoroughly tested, ensuring no downtime and simple rollback in the event of problems.

- **Rolling deployment**: In this method, the old version is gradually replaced in order to keep at least a portion of the system operational throughout the update.

Imagine you are an ETL data engineer in charge of a pipeline that integrates raw data into a data warehouse. To include new data sources, the pipeline needs to be updated frequently. You could set up a CI/CD pipeline using tools for deployment automation that automatically tests new code for data extraction and transformation before deploying it into the production environment once all tests are passed. By using a blue-green deployment strategy, you can easily roll back to an earlier version if you find any data inconsistencies or problems, maintaining data integrity and system dependability.

In any contemporary data engineering environment, automated deployment is not just a *nice-to-have* feature but rather a necessity. It reduces errors, frees up your time, and lets you concentrate on system improvement rather than deployment firefighting.

After exploring the workings and significance of deployment automation, let's turn our attention to the third and final pillar of automation in data engineering: monitoring. This section will go over how monitoring keeps your automated systems operating as they should while also offering useful information for future improvements.

Monitoring

Monitoring is a crucial part of the CI/CD pipeline that is frequently disregarded but is essential for preserving the functionality, performance, and dependability of your systems. Even the most advanced automated systems can malfunction in the absence of sufficient monitoring, resulting in data loss, downtime, or worse. Monitoring offers in-the-moment insights into a range of metrics and performance indicators, enabling you to take proactive measures to address problems before they become serious ones.

There are, primarily, three types of monitoring in the field of data engineering:

- **Application monitoring**: Monitoring your applications' performance and availability includes keeping track of their data pipelines, APIs, and other services as well as their performance. Typically, metrics such as throughput, error rates, and response time are tracked.

- **Infrastructure monitoring**: Monitoring the health of the underlying systems that support your data pipelines and applications is what is meant by infrastructure monitoring. This may include network latency, disk I/O, CPU usage, and memory usage.

- **Log monitoring**: This involves gathering, analyzing, and visualizing log files produced by your infrastructure and applications. Specific errors or trends that may not be picked up by other monitoring techniques can be found with the aid of log monitoring.

You have access to a number of tools to implement efficient monitoring. Grafana is frequently used to visualize those metrics in real-time dashboards, while Prometheus is a potent open source tool for gathering metrics and configuring alerts. They work together to offer a reliable monitoring option that can be customized to meet the specific requirements of your data engineering projects.

Best practices include creating dashboards that effectively communicate the health and performance of your systems, configuring alerts that are useful and alert-fatigue-free, and routinely reviewing your monitoring setup to adjust to changing needs or difficulties.

Now that you have a firm understanding of the role that monitoring plays in automation, let's move on to the next crucial competency in your data engineering toolkit: mastering Git and version control, the foundation of coordinated and collaborative code management.

Mastering Git and version control

Code is a dynamic entity that is constantly being improved by numerous contributors and deployed across a variety of environments in the world of software and data engineering. The choreography that keeps this complex dance of code development coordinated and manageable is Git and version control. This section aims to provide you with the necessary information and best practices for effectively using Git and version control systems. You'll discover how to keep track of changes, cooperate with team members, control code branches, and keep a record of the development of your project.

Understanding Git and version control is essential for ensuring code quality, promoting collaboration, and avoiding conflicts, whether you're working on a small team or contributing to a significant data engineering project. Let's get started with the fundamental ideas and methods that will enable you to master this crucial facet of contemporary data engineering.

Git architecture and workflow

Multiple contributors can work on the same project at once thanks to the distributed version control system known as Git. Each developer in Git has a local repository that contains an exact replica of the project's history. Local commits are followed by pushes to a central repository, which is typically housed on a platform such as GitHub or GitLab. As merging and conflict resolution are fundamental components of the Git workflow, this enables collaboration without causing constant overwriting of one another's work.

Let's examine a few of the fundamental Git commands you'll frequently use:

- `Clone`: You can copy an existing repository to your local machine using the `[Repository_url] git clone` command.
- `Commit`: When you commit, your changes are saved to the local repository. Git add `[file_names]` is used to stage changes prior to committing, and `git commit -m Commit message` is used to commit the staged changes.
- `Push`: This sends your committed changes to your remote repository. GitHub push origin `[branch_name]`.
- `Pull`: This updates your local repository by downloading updates from a remote repository to your computer. `[Branch_name] git pull` origin.

After laying a solid foundation in the fundamentals of Git, let's move on to the next subject: exploring branches, merges, and best practices for efficiently managing challenging data engineering projects.

Branching and merging

Branching strategies are essential in data engineering projects because they enable multiple team members to work on distinct project components simultaneously without interfering with one another. Effective branching keeps the *master* branch stable, making it simpler to track changes, revert to earlier versions, and isolate features or bug fixes for specialized testing and deployment.

There are two common branching models that you might see:

- **Git-flow**: In this model, in addition to the main branch, there are separate branches for features, releases, and hotfixes. This gives projects with set release cycles a structured workflow they can use.

- **GitHub-flow**: Changes are made in feature branches and then immediately merged into the main branch, which is then deployed according to the GitHub-flow process. It is more adaptable and works with CD projects.

Taking the modifications from one branch and applying them to another is known as merging in Git. Usually, the `git merge` command is employed for this task. Conflicts arise when the identical lines in identical files are altered in both of the branches you are attempting to merge. These conflicts will be noted by Git, and they must be manually resolved before the merge operation is finished. This entails staging the resolved files before committing them and then editing the conflicting files to decide which changes to keep.

Let's look at an instance where you're tasked with adding a new data validation feature to an already existing ETL pipeline as a data engineer. If your team uses the Git workflow, you will branch off the `develop` branch and work on your changes in a new branch called `feature`. When your `feature` branch is finished, you will start a `pull` request to merge it back into the `develop` branch. Eventually, this will be merged into the `main` branch as part of a planned release.

Consider a different scenario in which you and another engineer are tasked with improving various components of an algorithm for processing data. You both make adjustments to different branches. Git flags a conflict when merging these branches because both of you changed the same method. To fix this, you would manually select which changes to keep in the conflicting file, possibly integrating both sets of optimizations, before successfully completing the merge.

The foundation of collaborative data engineering is an understanding of branching and merging. These procedures enable agile development, reliable versioning, and efficient teamwork—all of which are essential to the accomplishment of any project involving data engineering.

Collaboration and code reviews

Git collaboration involves more than just pushing your changes; it also involves staying in sync with other people. You can keep your local repository updated with the most recent changes from your team by using the `git pull` and `git fetch` commands. `git fetch` gives you an additional level of control by allowing you to review the changes before merging, in contrast to `git pull`, which will fetch the changes and immediately merge them into your current branch.

Data engineering is no different from other software development life cycles with regard to the importance of code reviews. Code reviews make sure that the code follows best practices and is both functional and compliant. They act as a platform for knowledge exchange and have the ability to identify potential problems before they are incorporated into the production code.

Data pipelines require version control just like code does. Each change should be recorded, regardless of whether it involves adding a new feature, fixing a bug, or improving performance. By providing an audit trail and making it simpler to roll back to earlier states, versioning your data pipelines ensures that you can quickly correct any mistakes or unintended effects.

Bonus tools to make working with Git easier

While Git's command-line interface is powerful, several tools can make your life easier:

- **Lazygit**: This is a Git terminal-based UI that makes common operations such as staging, committing, branching, and merging simpler. For those who prefer visual interactions, it offers a simple interface.

- **GitKraken**: This is a well-known cross-platform Git GUI that shows branches, commits, and merges in visual form. In order to facilitate collaboration, it also integrates with GitHub, GitLab, and Bitbucket.

- **SourceTree**: This is another user friendly Git GUI that clearly displays the history, branches, and commits of your repository. It is compatible with both Windows and macOS.

- **GitExtensions**: This is a Git GUI for Windows that provides tools for managing repositories, context menus for Windows Explorer, and simple branching and merging.

- **Fork**: This is a Windows and macOS Git client with a stylish user interface for managing Git repositories. It makes merging, branching, and resolving conflicts simpler.

- **Tower**: This is a powerful GUI for managing repositories that is offered by a Git client for macOS and Windows. It has capabilities such as integration with well-known Git platforms and the visualization of code history.

- **GitUp**: This is a macOS-only tool that provides a special way to view and work with Git repositories. It offers a graphic graph of the history of your repository.

- **Magit**: Magit is a robust Git interface that seamlessly integrates into the text editor, allowing you to access Git functionality without ever leaving Emacs, if you prefer working inside of it.

- **Git-fuzzy**: This is a device that makes it simple to quickly find and switch between branches using fuzzy search. When you have a large number of units it is advantageous.

- **GitAlias**: Although it is not a GUI tool, GitAlias enables you to make unique Git aliases to streamline and condense frequently used commands. Your workflow could be significantly accelerated by this.

Git workflows must frequently be adapted in data engineering to manage data artifacts, configuration files, and even data models. It is essential that all components of your data pipeline are versioned and tracked, not just the code. This may necessitate utilizing **Git Large File Storage** (**LFS**) for large data files or establishing a separate repository for configuration management. Integrating Git into your data engineering workflow so that both code and data artifacts are versioned, making it easier to roll back changes, audit, and collaborate, is crucial.

Imagine you are a member of a distributed team in charge of a sizable data pipeline. A global teammate of yours has made some improvements and pushed them to the remote repository. Before merging these changes into your local branch for additional testing, you can use `git fetch` to verify them.

Using another scenario, let's say you've significantly altered the way a key ETL process in your data pipeline runs. You open a `pull` request rather than just merging this into the main branch. Your coworkers review the code and make suggestions for enhancements, and when everyone on the team agrees, the changes are merged and versioned. Use GitKraken to visually track changes across branches for complicated tasks, or Lazygit to condense common Git commands for quicker processing.

The importance of effective teamwork and thorough code reviews cannot be understated in the field of data engineering, where multiple people frequently collaborate on intricate, interdependent systems. This collaboration can be smooth and effective by properly using Git and supporting tools.

Now that we've reviewed the fundamentals of Git, we will focus our attention on data quality monitoring.

Understanding data quality monitoring

Equally important as the efficiency of your pipelines in data engineering is the quality of your data. Inaccurate analyses, flawed business decisions, and a loss of faith in data systems can result from poor data quality. Monitoring data quality is not just a one-time activity but a continuous process that needs to be integrated into your data pipelines. It ensures that the data ingested from various sources conforms to your organization's quality standards, thereby ensuring that the insights derived are trustworthy and actionable.

Data quality metrics

In data engineering, the quality of your data is just as essential as the efficacy of your pipelines. Poor data quality can result in erroneous analyses, faulty business decisions, and a loss of confidence in data systems.

Setting up alerts and notifications

Not only does automation extend to monitoring, but also to alerting. The next step after configuring data quality checks is to configure alerts and notifications. This can be done through various channels, such as email, Slack, or even automated scripts that trigger remedial actions. The objective is to notify the appropriate parties as soon as a data quality issue is identified, allowing for prompt remediation.

Imagine an online retailer that relies on real-time analytics to make inventory decisions. A minor data error could result in overstocking or understocking, both of which are costly errors. By implementing real-time data quality monitoring, the company is able to detect data discrepancies as they occur, enabling immediate corrective action. This case study illustrates the importance of data quality monitoring to operational effectiveness and decision-making.

Now that we have tackled how to monitor data quality, we will proceed to understand pipeline catch-up and recovery techniques.

Pipeline catch-up and recovery

In the world of data engineering, failure is not a question of *if* but *when*. Data pipeline failures are inevitable, regardless of whether they are caused by server outages, network problems, or code bugs. The ability to recover from these failures is what differentiates a well-designed pipeline from a fragile one. Understanding the types of failures that can occur and their potential impact on your pipeline is the first step in designing a resilient system.

Through a combination of redundancy, fault tolerance, and quick recovery mechanisms, data pipelines achieve resilience. Redundancy is the presence of backup systems in the event of a system failure. Fault tolerance is the process of designing a pipeline to continue operating, albeit at a reduced capacity, even if some components fail. Quick recovery mechanisms, on the other hand, ensure that the system can resume full operation as quickly as possible following a failure.

When a data pipeline fails, there is typically a backlog of data that must be processed once the system is operational again. Here, catch-up strategies become relevant. Parallel processing, batch processing, and data prioritization can assist your pipeline in catching up quickly. During the catch-up phase, you may prioritize real-time analytics data over historical data, for instance. The choice of strategy will depend on the pipeline's particular requirements and constraints.

When a failure is detected, recovery mechanisms are the predetermined procedures and automated scripts that are activated. These may range from simple restarts to complex workflows with multiple steps, such as data validation, rollback, and notifications. Implementing effective recovery mechanisms is essential for minimizing downtime and ensuring data integrity.

Having delved into the resilience of data pipelines, let's turn our attention to the final piece of the puzzle: implementing CD. The next section will guide you through the best practices for automating your data pipeline deployments, ensuring that your code changes are safely and efficiently transitioned into production.

Implementing CD

The capacity to release changes quickly and reliably is not just a luxury in the rapidly changing field of data engineering, but rather a necessity. The technique that fills this need and serves as the cornerstone of contemporary DevOps practices is CD. The practical aspects of CD will be covered in this section, with a special emphasis on crucial elements such as deployment pipelines and the use of infrastructure as code.

The goal of CD is to completely automate the transfer of code changes from development to production, minimizing the need for manual intervention and lowering the possibility of human error. Data engineers can more effectively handle tasks ranging from minor updates to significant features by utilizing CD, and they can also make sure that code quality is maintained throughout all environments. You will learn more about deploying dependable and strong data pipelines, managing infrastructure, and achieving a high level of automation in your data engineering workflows as you progress through this section. The following screenshot depicts the overall steps of CI/CD:

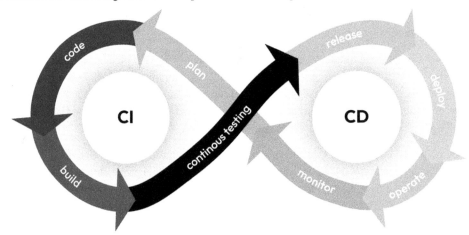

Figure 14.1 – CI/CD life cycle

Let's examine the procedures and tactics for integrating CD into your data engineering practices.

Deployment pipelines

Deployment pipelines allow code to move from a developer's local machine to a production environment in a controlled, transparent, and automated manner. They act as the automated manifestation of your deployment process. In the field of data engineering, these pipelines are essential because they control the deployment of ETL jobs, data transformations, and database changes, among other things.

There are several stages in a typical deployment pipeline, frequently including the following:

- *Code commit*: When new code is pushed to the repository, a code commit begins
- *Build*: This runs fundamental tests on the code and packages the application
- *Automated testing*: This executes a number of automated tests to ensure that the code is error-free and satisfies all specifications
- *Staging*: Code deployment to a staging environment allows for additional testing and validation
- *Deployment*: The code is deployed to production if all earlier steps are successful

Deployment pipeline creation and management are made easier by a variety of tools. Some of the well-known ones are Azure Pipelines, Jenkins, and GitLab CI/CD. These tools frequently offer a wealth of customization options to suit your particular deployment requirements and seamlessly integrate with existing code repositories.

Imagine you are a data engineer in charge of a data pipeline that compiles information from various sources into a single data warehouse. You've included a module for real-time data ingestion as part of a new feature. The deployment pipeline starts working the moment your code is committed. Initial unit tests are run after your code has been built. After those are passed, it advances to more in-depth automated tests to ensure that the real-time ingestion won't break any features already in use. Your changes are automatically deployed to the production data pipeline after being successfully validated in the staging environment and without the need for any manual intervention. Your new feature will be reliable, thoroughly tested, and integrated into the current system without any issues, thanks to the entire process.

Any data engineer who wants to succeed in a contemporary, agile, and automated environment must master the design and application of deployment pipelines. With pipelines, you can make sure that every code change is reversible in case problems arise and that it reaches the production environment in a stable and reliable manner.

Following our exploration of deployment pipelines, let's move on to **infrastructure as code** (IaC), another tenet of CD that gives your deployment environments programmability and version control.

Infrastructure as code

Instead of using interactive configuration tools or manual procedures, IaC refers to the practice of managing and provisioning computing infrastructure through machine-readable definition files. IaC can be used to set up and manage databases, data pipelines, and even sophisticated distributed systems such as Hadoop or Spark clusters in the context of data engineering. The benefit of IaC is that it extends the version control and automated deployment advantages that developers have enjoyed with application code to infrastructure.

To support IaC, several tools have been developed, each with advantages and disadvantages:

- *Terraform*: Known for its approach that is independent of the cloud, Terraform enables you to specify and provision infrastructure using a declarative configuration language. It is very extensible and works with a variety of providers.

- *Ansible*: Unlike Terraform, Ansible is agentless, which means your target machine does not need to run a daemon for it to function. Its playbook language, YAML, is used, and it excels at configuration management tasks.

IaC adopts a programmatic approach to infrastructure management, ensuring that resources are provisioned and managed in a predictable, standardized way. In data engineering, where the infrastructure is frequently as intricate and crucial as the data pipelines it supports, this becomes incredibly valuable.

Technical interview questions

You might be curious as to how important concepts and techniques such as automation, Git, and CD translate into the interviewing process after delving deeply into these topics. By emphasizing the types of technical questions you might be asked during a data engineering interview, this section aims to close that gap.

These questions aren't just theoretical; they're also meant to gauge your problem-solving skills and practical knowledge. Simple queries about SQL and data modeling will be covered, as well as more complicated scenarios involving distributed data systems and real-time data pipelines. The objective is to give you the tools you need to successfully respond to the countless questions that might be directed at you.

Now, let's look at the types of questions you might encounter and the best strategies for answering them:

Automation concepts:

- *Question 1*: What is the role of automation in CI/CD?

  Answer: Automation is at the core of CI/CD, which automates tasks such as building, testing, and deploying code changes. It ensures efficiency and consistency and reduces human error.

- *Question 2*: How does automation contribute to faster development cycles in CI/CD?

  Answer: Automation accelerates processes such as code testing and deployment, enabling rapid iteration and quicker feedback loops, and speeding up development cycles.

Git and version control:

- *Question 1*: Why is version control crucial to a CI/CD workflow?

  Answer: Version control tracks changes, enabling collaboration and providing a history of modifications. This history aids in debugging, code reviews, and maintaining code quality.

- *Question 2*: How does branching in Git facilitate CI/CD pipeline collaboration?

  Answer: Branching allows multiple developers to work on different features concurrently. Each branch can undergo testing before merging, ensuring stability in the main code base.

CD concepts:

- *Question 1*: What is the fundamental goal of CD?

  Answer: The primary aim of CD is to deliver code changes to production automatically and frequently, ensuring that new features and fixes are rapidly available to users.

- *Question 2*: How does a deployment pipeline contribute to reliable software releases?

  Answer: A deployment pipeline automates the testing, approval, and deployment stages. It ensures that code changes undergo consistent and thorough testing before reaching production.

Infrastructure as code:

- *Question 1*: What is the concept of IaC in CD?

  Answer: IaC involves scripting the provisioning of infrastructure using code. It ensures consistent and reproducible infrastructure setups, reducing manual configuration errors.

- *Question 2*: How can tools such as Terraform contribute to successful CD?

  Answer: Tools such as Terraform allow you to define infrastructure configurations as code, enabling easy replication of environments and ensuring that infrastructure changes are versioned.

Monitoring and rollbacks:

- *Question 1*: Why is monitoring essential to a CD environment?

  Answer: Monitoring tracks the performance and health of deployed applications. Automated alerts allow teams to detect issues promptly, enabling timely responses and maintenance.

- *Question 2*: How do rollbacks enhance the reliability of CD?

  Answer: Rollbacks provide a safety net in case of deployment failures. They enable quickly reverting to a previous version, minimizing downtime and maintaining system stability.

Deployment tools and efficiency:

- *Question 1*: How can deployment tools such as Jenkins contribute to a CI/CD workflow?

  Answer: Jenkins automates building, testing, and deploying code changes. It integrates with version control systems and streamlines the entire CI/CD process.

- *Question 2*: What is the role of Lazygit in a Git-based workflow?

  Answer: Lazygit provides a user friendly, terminal-based interface for Git interactions, making tasks such as branching, committing, and merging more efficient.

CD benefits:

- *Question 1*: How does CD improve collaboration among development and operations teams?

  Answer: CD fosters collaboration by automating processes and providing a shared understanding of code changes, ensuring that both teams work harmoniously.

- *Question 2*: What impact does CD have on software quality assurance?

  Answer: CD emphasizes automated testing, which improves code quality by catching bugs early. This results in higher-quality releases and reduced post-deployment issues.

Challenges and best practices:

- *Question 1*: What challenges are associated with CD, and how can they be mitigated?

  Answer: Challenges can include handling complex deployment scenarios and ensuring consistent environments. Mitigation involves thorough testing, proper version control, and well-defined deployment scripts.

These sample interview questions cover a variety of topics, including deployment tools, automation, version control, and best practices. Knowing about these ideas will give you the confidence to talk about CI and CD in interviews.

Summary

This chapter covered three fundamental data engineering topics: Git and version control, data quality monitoring, and pipeline catch-up and recovery techniques. We began by covering the fundamentals of Git, focusing on its role in team collaboration and code management. The importance of continuously monitoring data quality was then discussed, along with key metrics and automated tools. Finally, we addressed the inevitability of pipeline failures and provided strategies for resilience and speedy recovery.

Now that you have a solid grasp of continuous improvement techniques, it's time to move on to a subject that is essential in today's data-driven world: data security and privacy. We'll cover how to safeguard data assets, adhere to rules, and foster trust in the chapter that follows, all while making sure that data is available and usable for appropriate purposes.

15

Data Security and Privacy

Scale, efficiency, design, and (possibly most importantly) security and privacy all play important roles in navigating the world of data engineering. These components govern how safely and responsibly data is handled, not just as additional layers on top of the existing data landscape. Knowing how to secure and privatize this information is essential, whether you're working with confidential customer information, top-secret corporate documents, or even just operational data.

Security is a primary concern that must be integrated starting with the design phase and continuing through deployment and maintenance in the world of data engineering. Just as data privacy used to be a *nice-to-have*, it is now a *must-have*, thanks to regulations such as the **General Data Protection Regulation (GDPR)** and the **California Consumer Privacy Act (CCPA)**.

We want you to leave this chapter with a firm grasp of the fundamental ideas and procedures underlying data security and privacy. In addition to being necessary for daily tasks, these abilities are frequently assessed during data engineering interviews. This chapter aims to prepare you to be an expert in ensuring security and compliance by teaching you the fundamentals of data access control and delving deeply into the mechanics of anonymization and encryption. We'll also discuss the fundamentals of keeping systems up to date to prevent vulnerabilities.

These are the skills you will learn in this chapter:

- Grasping the importance and techniques of who has access to which data
- Respecting user privacy and regulatory compliance by learning how to de-identify data so that individuals can no longer be easily identified

In this chapter, we're going to cover the following main topics:

- Understanding data access control
- Mastering anonymization
- Applying encryption methods
- Foundations of maintenance and system updates

Understanding data access control

Data is frequently referred to as the new oil in the contemporary digital ecosystem because it is a crucial resource that powers businesses and influences decision-making. Data, however, is much simpler to access, copy, and spread—sometimes even unintentionally—than oil. The significant security risk posed by this accessibility calls for a strong framework to regulate who can access which data and in which circumstances.

That essential framework is called **data access control**. It establishes the limits within your data architecture, deciding how and with whom certain pieces of information may be interacted. Without strict access controls, there is an exponentially greater chance that sensitive information will end up in the wrong hands. Along with financial losses, this could have serious legal repercussions, especially in light of the present-day strict laws governing data protection.

We will delve deeply into the details of data access control in the following subsections. We'll start by comprehending the basic categories of access levels and permissions that specify the range of interactions a user can have with data, such as read, write, and execute. Next, we'll look at the crucial difference between authentication and authorization, two words that are frequently used interchangeably but have different meanings when it comes to access control. Then, we'll discuss **role-based access control** (**RBAC**), a potent method for structuring and controlling user access according to roles within an organization. Lastly, we'll look at **access control lists** (**ACLs**), a more detailed technique for establishing precise data access policies.

Not only is it a best practice, but it's also essential to understand data access control. Having a thorough understanding of these principles will not only make you a more capable data engineer but also a custodian of data safety, regardless of whether you're managing large data lakes, putting **Extract, Transform, Load** (**ETL**) pipelines into place, or even just querying a database. So, let's get started.

Access levels and permissions

The foundation for defining user interactions with data in the context of data security are three core access levels: **read**, **write**, and **execute**. A user with *read* permission can view the data but not change it. For analytical roles that require data inspection but not alteration, this is frequently crucial. *Write* permission takes things a step further by enabling users to edit the data, whether that means adding new records, updating old ones, or deleting data. *Execute* permission, which enables users to run particular programs that might change data or the state of the system, is used in the context of executable files and scripts.

Depending on the requirements of the organization and the sensitivity of the data, the scope of these permissions can range from being very general to being very specific. Giving a specific role within the company read access to a whole database may be a part of general access. If the data is sensitive, this broad level of access may be risky but is simpler to manage. Granular access, on the other hand, regulates permissions at a much finer level—possibly down to particular tables, rows, or even particular attributes within a database. Granular permissions give you a tighter hold on data access, but as the number of users and data objects increases, they can be difficult to manage. Considering the particular data being handled, its sensitivity, and the particular needs of the roles needing access will help you decide between granular and general access.

Building a secure data environment begins with understanding these basic concepts of access levels and permissions. Let's now look at the crucial differences between authentication and authorization, the two fundamental building blocks that support these access controls.

Authentication versus authorization

Although *authentication* and *authorization* are frequently used synonymously, they have different functions in the context of data security. Verifying a user, system, or application's identity is the process of authentication. It provides a response to the query, "Are you who you say you are?" On the other hand, authorization takes place following authentication and decides which resources the authenticated user has access to. It provides an answer to the query, "What are you allowed to do?"

For authentication, a variety of techniques are available, each with a different level of security. The simplest form of authentication is password-based, which uses a username and password to confirm identity. Although easy to use, this technique may be susceptible to phishing attacks or brute-force attacks. Token-based authentication uses a token that is frequently created by the server and then supplied by the user for subsequent interactions, providing a stateless and frequently more secure approach. **Multi-factor authentication** (**MFA**) adds an extra layer of security by requiring two or more verification methods, such as a password, a mobile device, or even biometric information such as a user's fingerprints.

Let's think about a possible scenario for a data engineer. Imagine being in charge of a data lake that includes both sensitive financial data and general company records. For the purposes of their work, employees from various departments require access to certain parts of this data lake. In this case, you would first authenticate users to make sure they are in fact authorized employees, possibly using a combination of password-based and MFA methods. Once they had been verified, their authorization levels would have determined whether they could only read general records (possibly for data analysis) or write to private financial sections (for roles such as financial analysts or upper management).

Setting up a reliable access control system requires an understanding of the subtle differences between authentication and authorization. Your organization's data can be protected with a layered security protocol when combined with specific access levels and permissions. Let's continue by looking at RBAC, a methodical method for managing this complicated landscape of authorizations and permissions.

RBAC

RBAC provides a structured method for controlling user access to various resources within an organization. The fundamental idea behind RBAC is fairly simple: users are assigned roles to segregate permission levels. In large or complex environments, this framework makes it simpler to manage, scale, and audit access permissions.

The definition and identification of roles within the organization is the first step in the implementation of RBAC. These roles ought to be compatible with the users duties and job functions. *Data analyst*, *database administrator*, and *financial officer* roles, for instance, each have a unique set of privileges suited to their requirements. The authorized interactions (read, write, and execute) that the roles can have with particular resources are referred to in this context as **privileges**. RBAC's flexibility allows roles to be easily updated, expanded, or contracted as organizational needs change.

Let's go over an instance in which a data engineer might find themselves. Imagine you are in charge of a multi-tenant database that provides access to various **business units** (**BUs**) such as marketing, sales, and finance. Due to the sensitivity of the data and the need for departmental privacy, only certain tables in this database should be accessible to each department. You can create roles such as `Marketing_User`, `Sales_User`, and `Finance_User` and assign each one the proper permissions using RBAC. Then, you assign the appropriate role to each user in those departments, ensuring that they have secure and useful access to the database.

Understanding how to maintain a harmonious balance between operational functionality and data security is made possible by RBAC. Let's explore ACLs, another layer that enables even finer-grained definitions of data access policies, with this basic knowledge in hand.

Implementing ACLs

When it comes to specifying access policies for resources, ACLs provide an additional level of granularity. An ACL is essentially a table listing individual entities and their particular permissions with respect to a given resource. ACLs offer a way to set permissions at the user or entity level, enabling more precise control than RBAC, which bases permissions on roles.

ACLs typically come in two flavors: mandatory and discretionary. **Discretionary ACLs** (**DACLs**), which let the owner of the resource specify which users have what kind of access, are frequently used for resources such as files or databases. On the other hand, users cannot alter **mandatory ACLs** (**MACLs**), which are typically enforced by organizational policy or compliance requirements. DACLs are more adaptable, but as the number of users and permissions rises, they risk becoming complex. Although MACLs are more rigid, they are frequently necessary to maintain compliance with laws such as GDPR or the **Health Insurance Portability and Accountability Act** (**HIPAA**).

Consider the scenario where you are a data engineer in charge of a project involving a highly sensitive dataset that includes both client and employee data. Different stakeholders, such as customer relations, legal, and HR teams, require various levels of access to this dataset. You can use ACLs to specify that the customer relations team can read and write to the *client* table while the legal team can only read the *contracts* table. The HR team can read and write to the *employee* table. ACLs are a good option in this situation because this level of detail is hard to manage through RBAC alone.

Having examined the specifics of ACLs, you are now well equipped with a variety of methods for successfully controlling data access. Let's now turn our attention to methods for anonymizing data, a crucial component of data privacy.

Mastering anonymization

The privacy and security of data cannot be overstated in a world that is becoming more and more data-driven. Controlling who has access to data is a crucial component of data security, as we've already discussed. There are circumstances, though, in which sharing the data itself may be necessary for analytics, testing, or outside services. In these circumstances, merely restricting access is insufficient; the data must be transformed in a way that preserves its analytical value while protecting the identity of the individuals it represents. Techniques for anonymization are useful in this situation.

Sensitive information is shielded from being linked to particular people by anonymization, which acts as a strong barrier. Understanding data anonymization techniques has become essential for any data engineer in light of growing data privacy concerns and strict data protection laws such as GDPR and CCPA.

The following subsections will discuss different data anonymization strategies that can increase the efficiency and security of data sharing. We will explore, among other techniques, generalization, perturbation, and k-anonymity principles. We'll also look at useful strategies for hiding personal information that, if revealed, could seriously jeopardize someone's privacy.

You will learn invaluable skills as you proceed through the upcoming sections that are essential not only for protecting data but also for guaranteeing that your company stays in compliance with changing data privacy laws. Let's start by talking about the various data anonymization methods you can use.

Masking personal identifiers

The masking of personal identifiers takes a more specific approach to data privacy than anonymization techniques, which take a more general approach. **Personal identifiers** are specific pieces of data that can be used to identify a particular person, such as license plate numbers, email addresses, and social security numbers. When using raw data for analytics, testing, or development while protecting sensitive information, masking these identifiers is essential.

There are numerous techniques and equipment created for this. One straightforward method is to substitute a generic symbol, such as an asterisk (*) or X, for each character in a string. A social security number such as *123-45-6789*, for instance, could be disguised as *--6789. Format-preserving encryption is one of the more sophisticated techniques, where the content of the data is scrambled but the format of the data is preserved. This is especially helpful if the data needs to pass validation checks or be used in systems that demand a specific format.

Tools for masking personal identifiers include programs such as Data Masker and Delphix, and even built-in capabilities in databases such as SQL Server and Oracle. To help organizations comply with data protection laws, these tools can be set up to automatically detect and mask personal data.

Your strategy for data privacy and security will be strengthened by the ability to mask personal identifiers, which effectively complements more general data anonymization techniques. Next, we will examine data encryption as a way to add yet another layer of security.

Applying encryption methods

We've covered access control mechanisms and data anonymization techniques in our exploration of data security and privacy, both of which offer substantial layers of defense. What happens, though, if the data must be transmitted or stored securely but still be in its original, recognizable form for some operations? This is where encryption techniques are useful.

A data engineer's security toolkit's Swiss Army knife is encryption. Encryption techniques can guarantee that your data stays private and intact whether you're storing it at rest, sending it over a network, or offering a secure method for user authentication. The different types of encryption techniques, such as symmetric and asymmetric encryption, as well as more specialized protocols such as **Secure Sockets Layer (SSL)** and **Transport Layer Security (TLS)**, will be the focus of the next subsections.

Understanding how to manage and implement encryption is essential for both data security and regulatory compliance. As you read through the subsections, you will develop a solid understanding of the most important data encryption techniques and how to use them in a variety of situations that a data engineer might run into.

Starting off this section, let's examine the fundamentals of data encryption techniques and their various applications.

Encryption basics

There are two main categories to consider when encrypting data: symmetric encryption and asymmetric encryption.

The same key is used in **symmetric encryption** for both encryption and decryption. As a result, it is quick and effective, making it ideal for encrypting large amounts of data. However, managing and distributing the encryption key securely is difficult because if an unauthorized person obtains it, they can quickly decrypt the data.

On the other hand, **asymmetric encryption** employs a pair of keys, one for encryption and another for decryption. The private key, which is required for decryption, is kept secret, while the public key, which is used for encryption, can be shared publicly. Although more computationally intensive, this method is more secure and is therefore better suited for secure communications or identity verification than for large datasets.

In both symmetric and asymmetric encryption, **key management** is crucial. Key management issues can make the entire encryption process vulnerable. This frequently entails secure key rotation, key storage, and occasionally even key destruction procedures.

Let's take an instance that a data engineer might run into: Imagine that you oversee a system that deals with private user data such as financial transactions. Due to its effectiveness with large datasets, symmetric encryption is a good choice for securely storing this data. However, you might use asymmetric encryption to increase security when transmitting payment information between the client and server during a transaction.

A critical first step in securing your data is comprehending the fundamentals of encryption and key management. Next, we'll examine two more specialized protocols, SSL and TLS, which are frequently used for secure data transmission.

SSL and TLS

Secure network communications are made possible by the cryptographic protocols SSL and TLS. Although TLS is the successor to SSL, the term *SSL* is frequently used informally to refer to both. To maintain data confidentiality and integrity during transmission, these protocols combine symmetric and asymmetric encryption methods.

Installing a server certificate, which acts as a public key for creating encrypted communications, is typically required to implement SSL/TLS. The decryption process makes use of the server's private key, which is kept secret. Data transferred between the server and clients is encrypted once the certificate is in place, making it extremely secure against data tampering or eavesdropping.

Consider yourself a data engineer tasked with creating an ingestion pipeline for data from various external APIs. To maintain both data integrity and confidentiality, it is crucial to make sure that this data is transmitted securely. In this case, you would make sure that all communications between your server and the external APIs are encrypted using SSL/TLS. As a result, your server's data packets that are transferred to the APIs are encrypted, protecting them from unauthorized access or modification while in transit.

After discussing the significance and use of SSL and TLS, let's move on to the fundamental components of system updates and maintenance—a crucial but frequently disregarded area in the landscape of data security and privacy.

Foundations of maintenance and system updates

We've covered how to protect access to your data up to this point, as well as how to protect it while it's in transit and at rest. Even after these safeguards are in place, a data engineer must continue to work to ensure data security and privacy. Your data security infrastructure requires ongoing maintenance and regular system updates to adapt to new threats and compliance requirements, just as with a well-tuned engine.

Regular updates and version control

Regular system updates include minor fixes, significant upgrades, and new feature additions, and they go hand in hand with software patching. It's essential to have a clear schedule in place before putting these updates into action. Updates are first implemented in a development or testing environment before being rolled out in the production system, and this staged approach frequently works well. Here, **version control systems (VCSs)** can be extremely helpful because they let engineers keep track of changes, roll back to earlier versions if necessary, and work together more efficiently.

When operating on outdated systems, you miss out on enhanced features and optimizations. You're also more exposed to security risks. Outdated software might not adhere to current data protection laws, which could compromise data integrity and lead to costly fines and reputational harm.

Consider yourself a data engineer in charge of a big data platform that hasn't received an upgrade in a few years. The vendor has already made several new releases, each with enhanced features and security. You observe performance problems and an increase in failed security audits as you put off updating. Such a situation exposes your entire data ecosystem to risk, turning it into a ticking time bomb of security flaws and operational inefficiencies.

In conclusion, consistent updates and strong version control are crucial because they form the foundation of a strong and secure data infrastructure. The crucial procedures of monitoring, logging, and auditing will be covered in the next chapter to provide ongoing oversight and assurance for your data systems.

Summary

We covered key topics in data security and privacy in this chapter, including encryption techniques and access control mechanisms such as authentication and authorization. We also emphasized how important routine maintenance and system updates are for protecting data. Examples from the real world were given to help the data engineer understand these concepts.

As we proceed to the last chapter, we'll put your knowledge of these subjects to the test with a new set of interview questions, preparing you for both real-world problems and interviews.

16
Additional Interview Questions

Welcome to *Additional Technical Interview Questions*, an essential chapter on your path to becoming a successful data engineer. While the fundamentals, such as SQL queries, data modeling, and pipeline orchestration, were covered in earlier chapters, this chapter aims to introduce you to the more subtle, and frequently more difficult, aspects of data engineering interviews.

Many candidates perform admirably on the fundamental questions but struggle on the more difficult ones, which are frequently the ones they will encounter while working. To close those knowledge and readiness gaps, this chapter has been written. You can increase your confidence and show potential employers that you have the skills and knowledge necessary to handle real-world data engineering challenges by practicing these difficult questions and mastering the underlying concepts.

Now, let's take a look at these additional questions:

- *Question 1*: What is a data lake, and how does it differ from a data warehouse?

  Answer: A data lake is a storage repository that can hold vast amounts of raw data in its native format. Unlike a data warehouse, it allows flexible processing and analysis without predefined schemas.

- *Question 2*: Discuss the advantages and challenges of using cloud-based data storage solutions.

  Answer: Cloud-based storage provides scalability, accessibility, and cost-effectiveness. Nowadays, companies use multiple vendors to optimize their cloud performance and improve cloud flexibility.

- *Question 3*: Describe the concept of **change data capture** (**CDC**) in data engineering.

  Answer: CDC captures and tracks changes made to a database, enabling real-time synchronization between source and target systems.

- *Question 4*: What is data preprocessing, and why is it important in data analysis?

  Answer: Data preprocessing involves cleaning, transforming, and organizing raw data before analysis. It improves data quality and prepares data for accurate insights.

- *Question 5*: Discuss the advantages and limitations of batch processing and stream processing.

  Answer: Batch processing is suitable for high-volume data with delayed insights, while stream processing allows real-time analysis with lower latency but can be more complex to manage.

- *Question 6* What is the role of data engineering in enabling data-driven decision-making for organizations?

  Answer: Data engineering transforms raw data into usable information, facilitating accurate insights for decision-makers. It ensures that data is accessible, reliable, and timely for informed choices.

- *Question 7*: What is data versioning, and why is it essential in data engineering?

  Answer: Data versioning involves managing different iterations of datasets. It's crucial for maintaining historical records, tracking changes, and ensuring reproducibility.

- *Question 8*: Describe the concept of data masking and its importance in data security.

  Answer: Data masking involves replacing sensitive data with fictional data while retaining the original format. It's crucial for protecting sensitive information during testing and development.

- *Question 9*: What is data governance, and why is it important?

  Answer: Data governance involves establishing policies, roles, and responsibilities for data management. It's essential for ensuring data quality, compliance, and accountability.

- *Question 10*: Explain the concept of data anonymization and its relevance in data privacy.

  Answer: Data anonymization involves removing personally identifiable information from datasets to protect individual privacy while maintaining data utility for analysis.

- *Question 11*: How do you optimize a SQL query for performance?

  Answer: Query optimization involves using indexes, rewriting queries, and avoiding unnecessary joins to improve query execution speed.

- *Question 12*: Explain the concept of data lineage and its significance in auditing.

  Answer: Data lineage traces data movement through processes, ensuring transparency and accountability. It's vital for regulatory compliance and identifying data anomalies.

- *Question 13*: How do you handle schema changes in a database without causing disruption?

  Answer: Techniques such as blue-green deployment or feature flags can help manage schema changes without affecting users.

- *Question 14*: Describe the concept of **change data capture** (**CDC**) in data engineering.

  Answer: CDC captures, and tracks changes made to a database, enabling real-time synchronization between source and target systems.

- *Question 15*: What is data preprocessing, and why is it important in data analysis?

  Answer: Data preprocessing involves cleaning, transforming, and organizing raw data before analysis. It improves data quality and prepares data for accurate insights.

- *Question 16*: Discuss the challenges and benefits of working with unstructured data.

  Answer: Unstructured data needs a predefined structure. Challenges include parsing and analyzing such data, while benefits include extracting insights from diverse sources.

- *Question 17*: How does data serialization work, and what are its use cases?

  Answer: Data serialization converts data structures into a suitable format for storage or transmission. It's used to store objects in databases or transmit data over a network.

- *Question 18*: How do you ensure data security in a distributed system?

  Answer: Data security in distributed systems involves authentication, encryption, access control, and auditing mechanisms to prevent unauthorized access.

- *Question 19*: Explain the concept of data encryption and its importance in data protection.

  Answer: Data encryption involves transforming data into a secure format to prevent unauthorized access. It's crucial for protecting sensitive information.

- *Question 20*: Describe the role of data engineers in ensuring data privacy.

  Answer: Data engineers play a vital role in implementing privacy measures such as data masking, access controls, and anonymization to protect user information.

- *Question 21*: What are the challenges of working with real-time data processing?

  Answer: Challenges include low latency requirements, managing data streams, ensuring data accuracy, and handling potential bottlenecks in processing.

- *Question 22*: How do you perform version control for data pipelines?

  Answer: Version control for data pipelines involves using tools such as Git to manage changes to pipeline code, configurations, and dependencies.

- *Question 23*: Explain the importance of monitoring and alerting in data engineering.

  Answer: Monitoring tools track the performance of data pipelines and systems. Alerting mechanisms notify data engineers of anomalies or issues that require attention.

- *Question 24*: What are the benefits of using containerization in data engineering?

  Answer: Containerization, such as with Docker, provides isolation, portability, and consistency in deploying data engineering applications across different environments.

- *Question 25*: How do you optimize data processing workflows for performance?

  Answer: Optimization involves using parallel processing, caching, and efficient algorithms to enhance data processing speed.

- *Question 26*: Discuss the advantages and limitations of batch processing and stream processing.

  Answer: Batch processing is suitable for high-volume data with delayed insights, while stream processing allows real-time analysis with lower latency but can be more complex to manage.

- *Question 27*: How do you optimize data processing workflows for performance?

  Answer: Optimization involves using parallel processing, caching, and efficient algorithms to enhance data processing speed.

- *Question 28*: What is data versioning, and why is it essential in data engineering?

  Answer: Data versioning involves managing different iterations of datasets. It's crucial for maintaining historical records, tracking changes, and ensuring reproducibility.

- *Question 29*: What is data lineage, and why is it essential for regulatory compliance?

  Answer: Data lineage traces data flow and transformations, aiding auditing and regulatory compliance by showing how data is used and transformed.

- *Question 30*: Discuss the advantages of using NoSQL databases in specific data engineering scenarios.

  Answer: NoSQL databases provide flexibility, scalability, and better handling of unstructured data. They suit scenarios where schema flexibility and high write throughput are essential.

- *Question 31*: Describe the concept and advantages of data normalization.

  Answer: Data normalization eliminates data redundancy and anomalies by organizing data into separate tables. This reduces data duplication and enhances data consistency.

- *Question 32*: How do you ensure data security in data pipelines?

  Answer: Data security in data pipelines involves encryption, access controls, and monitoring mechanisms to protect data during transit and processing.

- *Question 33*: Explain the concept of **Extract, Transform, Load** (**ETL**) and its role in data engineering.

  Answer: ETL involves extracting data from source systems, transforming it into a suitable format, and loading it into a target system for analysis and reporting.

- *Question 34*: Describe the challenges of data migration between different storage systems.

  Answer: Challenges include data format conversion, data integrity preservation, and minimizing downtime during migration.

- *Question 35*: Explain the role of data modeling in data engineering.

  Answer: Data modeling involves creating representations of data structures, relationships, and constraints. It aids in database design and understanding data flow.

- *Question 36*: Describe the concept of data masking and its importance in data security.

  Answer: Data masking involves replacing sensitive data with fictional data while retaining the original format. It's crucial for protecting sensitive information during testing and development.

These interview questions cover a broad spectrum of data engineering topics and concepts, ensuring you're well prepared to engage in technical discussions during interviews. Understanding these principles and their practical applications will showcase your expertise in the field and enhance your confidence when addressing interview queries.

Index

Packtpub.com

Subscribe to our online digital library for full access to over 7,000 books and videos, as well as industry leading tools to help you plan your personal development and advance your career. For more information, please visit our website.

Why subscribe?

- Spend less time learning and more time coding with practical eBooks and Videos from over 4,000 industry professionals

- Improve your learning with Skill Plans built especially for you

- Get a free eBook or video every month

- Fully searchable for easy access to vital information

- Copy and paste, print, and bookmark content

Did you know that Packt offers eBook versions of every book published, with PDF and ePub files available? You can upgrade to the eBook version at packtpub.com and as a print book customer, you are entitled to a discount on the eBook copy. Get in touch with us at customercare@packtpub.com for more details.

At www.packtpub.com, you can also read a collection of free technical articles, sign up for a range of free newsletters, and receive exclusive discounts and offers on Packt books and eBooks.

Other Books You May Enjoy

If you enjoyed this book, you may be interested in these other books by Packt:

Data Wrangling with SQL

Raghav Kandarpa, Shivangi Saxena

ISBN: 978-1-83763-002-8

- Build time series models using data wrangling
- Discover data wrangling best practices as well as tips and tricks
- Find out how to use subqueries, window functions, CTEs, and aggregate functions
- Handle missing data, data types, date formats, and redundant data
- Build clean and efficient data models using data wrangling techniques
- Remove outliers and calculate standard deviation to gauge the skewness of data

SQL Query Design Patterns and Best Practices

Steve Hughes, Dennis Neer, Dr. Ram Babu Singh, Shabbir H. Mala, Leslie Andrews, Chi Zhang

ISBN: 978-1-83763-328-9

- Build efficient queries by reducing the data being returned
- Manipulate your data and format it for easier consumption
- Form common table expressions and window functions to solve complex business issues
- Understand the impact of SQL security on your results
- Understand and use query plans to optimize your queries
- Understand the impact of indexes on your query performance and design
- Work with data lake data and JSON in SQL queries
- Organize your queries using Jupyter notebooks

Packt is searching for authors like you

If you're interested in becoming an author for Packt, please visit `authors.packtpub.com` and apply today. We have worked with thousands of developers and tech professionals, just like you, to help them share their insight with the global tech community. You can make a general application, apply for a specific hot topic that we are recruiting an author for, or submit your own idea.

Share Your Thoughts

Now you've finished *Cracking the Data Engineering Interview*, we'd love to hear your thoughts! Scan the QR code below to go straight to the Amazon review page for this book and share your feedback or leave a review on the site that you purchased it from.

`https://packt.link/r/1-837-63077-1`

Your review is important to us and the tech community and will help us make sure we're delivering excellent quality content.

Download a free PDF copy of this book

Thanks for purchasing this book!

Do you like to read on the go but are unable to carry your print books everywhere?

Is your eBook purchase not compatible with the device of your choice?

Don't worry, now with every Packt book you get a DRM-free PDF version of that book at no cost.

Read anywhere, any place, on any device. Search, copy, and paste code from your favorite technical books directly into your application.

The perks don't stop there, you can get exclusive access to discounts, newsletters, and great free content in your inbox daily

Follow these simple steps to get the benefits:

1. Scan the QR code or visit the link below

https://packt.link/free-ebook/9781837630776

2. Submit your proof of purchase
3. That's it! We'll send your free PDF and other benefits to your email directly

www.ingramcontent.com/pod-product-compliance
Lightning Source LLC
LaVergne TN
LVHW080116070326
832902LV00015B/2617